Physics of Planetary Interiors

PHYSICS OF PLANETARY INTERIORS

Professor G H A Cole
University of Hull

Adam Hilger Ltd, Bristol

British Library Cataloguing in Publication Data

Cole, G. H. A.
Physics of planetary interiors.
1. Planets
I. Title
523.4 QB601

ISBN 0-85274-444-7 (hbk)
0-85274-445-5 (pbk)

Consultant Editor: **Professor A J Meadows**
Department of Astronomy, University of Leicester

Published by Adam Hilger Ltd, Techno House, Redcliffe Way, Bristol BS1 6NX.

Filmset by Mid-County Press, London SW15.
Printed in Great Britain by J W Arrowsmith Ltd, Bristol.

To T

Contents

INTRODUCTION

Many of the surface details of the planets known to the ancients (Mercury out to Saturn) are now familiar as the result of a wide variety of space mission encounters with the planets over the last two decades. New horizons will be opened when Voyager II encounters first Uranus (1986) and then Neptune (1995) on its journey out of the Solar System. Unfortunately, conditions in the interiors cannot be deduced directly from surface measurements alone and a wide range of arguments based on a theoretical understanding of the physics of matter under planetary conditions must be invoked if any reliable descriptions of the constitution of the planets and satellites are to be constructed. The development of such descriptions forms the subject of this book.

In collecting together a quantity of matter of planetary mass we produce a body whose gross properties are determined by the macroscopic average of the atomic behaviour. What can we expect the general properties of such a body to be? What are the special features which produce a planetary body with the properties of the type we observe in the Solar System? How is such a planet related to other possible celestial objects? These are the type of questions that will concern us in an attempt to discover the characteristic features of a planet in general terms.

Various astronomical approaches have led to the recognition of a common primitive cosmic abundance of the elements for various components of the Universe at large and it appears that the Solar System was formed from just such an abundance (or something very like it) 4.6×10^9 years ago. It is interesting to ask to what extent the observed properties of the planets and satellites relate to a composition of cosmic abundance: either in full or only partially? Possible compositions might involve particularly the main components of the abundance (for the major planets) or primarily the more rare components (for the terrestrial planets) perhaps involving also the less rare (as for the satellites of the major planets). Deduced deviations from the full cosmic abundance then raise the question of how the abundance became distorted in the case of the Solar System, but this must involve the speculative consideration of the formation of the System and will be excluded from our discussion.

This range of problems will be met, not by concentrating on individual planets and satellites, but by considering instead accumulations of material over a wide range of mass and a variety of constitutions. We study representative models of different types of planetary body even though each planet has unique features. The conditions are sought which best mirror various features of the different planets and satellites actually observed.

To do this, we concentrate on general principles of physics, and in particular consider the equilibrium that is possible between the force of universal gravitation pulling the constituent atoms together and the atomic forces which resist matter becoming concentrated beyond a characteristic amount. This involves the study of macroscopic matter under conditions of pressure and temperature which seem appropriate to planetary conditions. Larger accumulations of matter cause the atoms to be disrupted and other forces are then called into play to resist the inward pull of gravity to form an equilibrium arrangement. Degeneracy forces of electronic motion provide resistance to compression for some bodies (white dwarfs, mass < 1.44 Solar masses) and for stronger gravity, nuclear forces can be called into play (neutron stars). Even stronger gravity provides exotic alternatives; in these terms, a planetary body is recognised as a member of a hierarchy of related bodies, but of relatively low mass. Thermal forces play no part in these stable equilibria and in this sense these bodies are cold even though the temperatures may be high and the distribution within them non-uniform. In this respect, a conventional star presents a different and special transient equilibrium where thermal pressures oppose gravity while thermonuclear reactions can provide the necessary energy.

The constitution and composition to be expected of a body is controlled by its material mass to a remarkable degree, because this determines the strength of the gravitational force binding it together and which must be opposed by other forces. This is particularly the case for a planetary body and the compositions of large, medium and small planetary bodies (we shall be able to specify what is meant by these terms later) are found to be restricted in various ways. The detailed chemical composition of the planets is still unknown, although the general features are clear, and this uncertainty is accounted for naturally in our approach. We shall find it sufficient to invoke only the most general common features of matter for the study of the balance of the gravitational and the elastic forces of matter which controls the equilibrium of a planetary body, and in this way our arguments will have sufficient generality to apply to bodies that might equally be associated with any of a large category of stars other than the Sun. On this basis, we shall find quite generally that there is a maximum mass for a planetary body, beyond which it changes its character (in fact generally to become a star). Similarly, we shall be able to identify a natural lower limit to the mass (the smallest mass which will form a shape of symmetry) although the limit is not sharp. Referring to the members of the Solar System, the largest planet, Jupiter, is very little lighter than the maximum mass we calculate and the smaller icy satellites of Saturn are close to the lower limit. The satellites of Mars are below the limit. The other planets and main satellites spread across the range delineated by the largest and smallest bodies. These arguments are quite general and do not refer to the Solar System alone.

The physical properties of the members of the Solar System are set out in chapter 1, and the general equilibrium between gravitational and macroscopic elastic forces is explored in chapter 2, where the possible mass range for planetary structures is recognised. Consequences of the equilibrium are explored in chapter 3 and, in particular, consideration is given to the construction of a general equation of state for material under planetary conditions. The shape and internal mass distribution for a planetary body forms the subject of chapter 4.

Thermal energy is not a factor in determining the overall equilibrium of the planetary body and to this extent the body is cold. There are, however, internal heat sources, and the evolution of a body once formed is affected by thermal effects. Chapter 5 is devoted to comments on such thermal conditions. Magnetic fields of internal origin present interesting possibilities for exploring certain aspects of the interiors of planetary bodies, and these are considered in chapter 6.

The remaining chapters are devoted to wide-ranging surveys of the features that might be expected in bodies of different mass. It is here that some specific comparison is made between the results of our studies and the observed physical properties of the planets themselves. Chapter 7 is involved with a planet of large mass, and we have in mind the major planets here. Terrestrial type planets are considered in chapter 8, a grouping which includes also the Moon and Io as well as the inner planets. Finally, chapter 9 considers planetary bodies composed primarily of ice, and so is concerned with many of the satellites of the major planets.

The chapters are supported by auxiliary sections devoted to references and comments and these must be regarded as an integral part of the chapter with which they are associated. It is important to refer to them while the main chapter is read because they contain many additional aspects and special, selected details of the central themes of the main chapters. The selection of these reflects the prejudices of the author. No effort has been made to make the reference list in any way comprehensive. Rather, references have been selected which will themselves guide the reader to the wider literature. With this structure it is hoped the book will be of use to a wide range of readers and especially that it might convey something of the interest and excitement of the subject generally to a reader approaching it fresh for the first time. Having studied this book the reader should be the more able to appreciate the detailed literature for a particular planet in a wider context and in this way the present book is designed to be an introduction to the subject.

Finally, it is a pleasure to take this opportunity to acknowledge help in various ways with the writing of this book. By far the greatest is to my friend

Professor Michael Woolfson FRS, of the University of York. I have learned a great deal about planetary matters through many discussions with him over the years and it is a great pleasure to express my warm appreciation to him now. He also very kindly looked through the proofs for me and made a number of helpful comments. I am also much indebted to Professor A J Meadows of the University of Leicester for his careful reading of the manuscript early on; his comments and questions helped to improve the arguments at many points. Any shortcomings or errors in the book that remain cannot be ascribed to either gentleman; they are mine alone. I am lucky to have had Jim Revill and Ian Kingston of Adam Hilger Ltd working with me. They have given great support during the production of the book and have shown patience and understanding over extended deadlines. I offer them my best thanks. The excellent help of the other staff of Adam Hilger cannot go unmentioned. Comments and suggestions from readers to improve the arguments or eliminate errors of substance or accident that might remain will be most welcome.

1

SOME OBSERVATIONAL DATA

In this chapter we collect various data together describing the physical and orbital characteristics of the more important members of the Solar System. These data will form a background for our discussion.

1.1 PLANETARY MASSES AND RADII

The magnitudes of the masses and radii of the planets of the Solar System are listed in table 1.1. The Moon is included for reasons that will be clear later.

It is seen that the planets fall into two main groupings, and Pluto is different. The first group contains the four terrestrial planets (Mercury, Venus, Earth–Moon and Mars) with the Moon having the characteristics of a small planet (see figure 1.1). The second group consists of the four Jovian, or major, planets (Jupiter, Saturn, Uranus and Neptune). Pluto, at the edge of the Solar System, is a rather anomalous object as a planet, having a slightly smaller radius than the Moon and about one fifth the mass (see figure 1.1).

The planets have masses in the range 10^{22} kg to 2×10^{27} kg while their sizes fall within the range 10^3 to 10^7 km. The corresponding material densities range between 600 kg m^{-3} for Saturn (which has a density slightly more than half that of water) and about 5500 kg m^{-3} for the Earth (some ten times as great). We can express this range from an atomic point of view. The mass of a nucleon (a proton or neutron) is approximately 1.6×10^{-27} kg so the mass of a planet can be expressed alternatively in terms of the number of nucleons of which it is composed. From this we conclude that a planet contains somewhere between 10^{49} and 10^{54} nucleons, combined in various ways with electrons to form the range of atoms observed to constitute the Solar System.

Table 1.1 *The masses and the radii of the planets of the Solar System with the mass and radius of the Earth taken as unity in columns 3 and 5.*

Planet	Mass ($\times 10^{24}$ kg)	Mass (Earth = 1)	Equatorial radius ($\times 10^3$ km)	Radius (Earth = 1)	Mean density ($\times 10^3$ kg m^{-3})	Equatorial surface gravity (m s^{-1})
Mercury	0.330	0.056	2.44	0.382	5.42	3.78
Venus	4.87	0.815	6.05	0.949	5.25	8.60
Earth	5.976	1.00	6.38	1.000	5.52	9.78
Moon	0.0735	0.012	1.74	0.273	3.34	1.62
Mars	0.642	0.017	3.40	0.532	3.94	3.72
Jupiter	1.899×10^3	317.9	7.19×10	11.27	1.314	22.88
Saturn	5.686×10^2	95.15	6.00×10	9.44	0.690	9.05
Uranus	8.66×10	14.54	2.61×10	4.10	1.19	7.77
Neptune	1.030×10^2	17.23	2.475×10	3.88	1.66	11.00
Pluto	0.014	0.0023	1.44	0.226	1.14	0.451

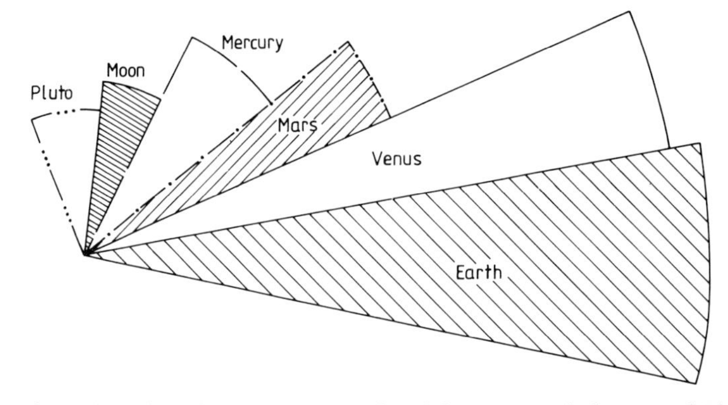

Figure 1.1 *The relative sizes, to scale, of the terrestrial planets and Pluto.*

The inferred densities listed in table 1.1 can be interpreted to give a preliminary indication of the chemical composition of the planets. It is immediately clear that the terrestrial planets must be deficient in the lighter elements in comparison with the Jovian planets.

1.2 COSMIC ABUNDANCES OF THE ELEMENTS

The relative abundances of the elements in the Galaxy can be inferred in a variety of ways involving a wide range of different techniques associated with

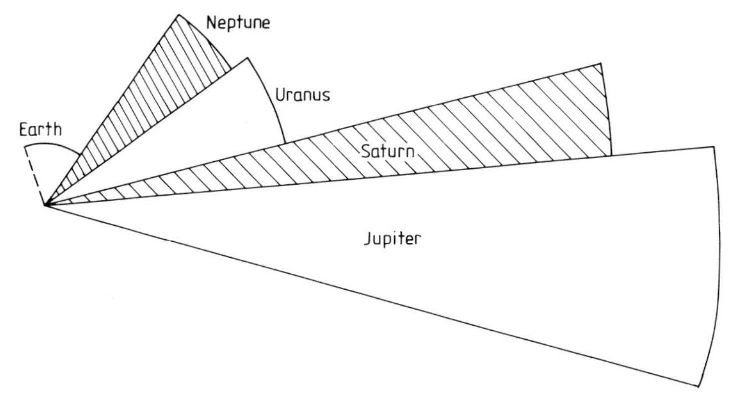

Figure 1.2 *The relative sizes, to scale, of the major planets.*

different disciplines. For the non-volatile materials a clue can be obtained from the study of surface material of the Earth and, as a consequence of the Apollo landings, also of the Moon. The study of the composition of meteorites includes material definitely known to be of non-terrestrial origin. The analysis of the solar wind provides information which can be interpreted in conjunction with the details of the solar spectrum, and includes volatile components.

The composition of cosmic rays also fits into this pattern. Stellar spectra and those of interstellar clouds give wider information. Ideas of cosmology and of the evolution of stars have allowed a theoretical understanding of the formation of the chemical elements through nuclear fusion processes and this has led to a quantitative assessment of the abundances. The result of all these varied analyses is contained in table 1.2.

It is significant that all these various approaches lead to a single abundance table. In very broad terms, the universal cosmic material is a mixture of about three hydrogen atoms to one helium atom, together with about one heavier atom per 3000 of the hydrogen–helium mixture. This does not conform to our experience on Earth because the Earth is a very unusual object.

The mixture of material listed in table 1.2 will be subject to considerable chemical activity during the formation and early evolution of the Solar System. Of special significance is the group (O, C, N) which is able to form a wide range of molecules such as CO, CO_2, N_2, O_2 and CN. In combination separately with hydrogen they form molecules such as H_2O, NH_3 and CH_4, among many others. These are all molecules well known in our usual environment. They are also the elements that appear to play so important a part in living material. More complicated material is formed by the association of these elements with those of less frequent occurrence such as magnesium, silicon and so on. Particularly important here are the compounds involving oxygen and silicon together with other elements, collectively called silicates. Silicates are widely distributed throughout the Earth and Moon; olivine ($(Mg, Fe)_2SiO_4$) and pyroxene ($(Ca, Mg, Fe)SiO_3$) are particularly abundant

Table 1.2 *The relative abundances of the elements by number and by mass with the data for silicon taken as standard.*

Element	Relative abundance by number (silicon = 1)	Relative abundance by mass
Hydrogen	3.18×10^4	0.9800 (Hydrogen, Helium)
Helium	2.21×10^3	
Oxygen	2.21×10	0.0133 (Oxygen, Carbon, Nitrogen)
Carbon	1.18×10	
Nitrogen	3.64	
Neon	3.44	0.0017
Magnesium	1.06	0.00365 (Magnesium to Nickel)
Silicon	1	
Aluminium	8.5×10^{-1}	
Iron	8.3×10^{-1}	
Sulphur	5×10^{-1}	
Calcium	7.2×10^{-2}	
Sodium	6×10^{-2}	
Nickel	4.8×10^{-2}	

and the various rearrangements due to elevated pressure lead to a complicated but attractive mineralogy. The condensed phases (whether solid or liquid) are particularly relevant for planetary interiors and the various densities are interesting from our present point of view. Silicates at zero pressure have densities according to composition in the general range 2600 to 3600 $kg\,m^{-3}$ although the upper limit can be higher if ferrous components are present. Pure iron has a density of about 7800 $kg\ m^{-2}$ at zero pressure, but this may be reduced by the presence of other elements such as sulphur. This probably represents the maximum magnitude for the density as far as planetary materials are concerned. At the lower end, water-ice has a density of about 920 $kg\,m^{-3}$ at low pressures (when the ice has the phase 1) but the density increases with pressure as the phase of the ice changes (see chapter 9) approaching a magnitude at high pressure of about 1500 $kg\,m^{-3}$. The densities of methane-ice and ammonia-ice are about half this magnitude; the density of methane-ice is about 528 $kg\ m^{-3}$. The rheology of ices is not very well understood at the present time. These various estimates refer to low pressures and could well be increased by as much as a factor two under the compression to be expected in a deep planetary interior.

We can, then, make a preliminary interpretation of the planetary densities listed in table 1.1 in terms of possible mineral materials. These are collected in table 1.3 and shown diagrammatically in figure 1.3. The densities of the terrestrial planets suggest a mixed composition of silicates and free ferrous

Table 1.3 *The mean densities and possible compositions of the planets of the Solar System.*

Planet	Mean density ($\times 10^3$ kg m^{-3})	Possible material composition
Mercury	5.42	Mainly iron with silicates.
Venus	5.25	Silicates with free ferrous materials comprising some 28% of the total mass.
Earth	5.52	Silicates with free ferrous materials comprising some 32% of the total mass.
Moon	3.34	Silicates with very little free ferrous material if any.
Mars	3.94	Silicates, possibly with some free ferrous materials.
Jupiter	1.314	Mainly hydrogen and helium but with heavier elements rather in excess of the cosmic abundance.
Saturn	0.690	Mainly hydrogen and helium but with heavier elements somewhat in excess of the cosmic abundance.
Uranus	1.19	Mainly hydrogen and helium but some water, carbon dioxide, ammonia, methane and some heavier elements.
Neptune	1.66	Water, carbon dioxide, ammonia and methane with hydrogen and helium.
Pluto	1.14	Mainly water-ice with silicates.

materials. The proportion of iron in Mercury will need to be high to account for the large density of so small an object, probably as high as 80%. For comparison, the pure ferrous content of the Earth is likely to be no more than about 32%. Of all the planets, Jupiter and Saturn alone are likely to have a composition close to the cosmic abundance and so can be considered together. Uranus and Neptune are different in that they contain a greater proportion of oxygen, carbon and nitrogen compounds. The terrestrial planets must have a very different composition, being deficient particularly in hydrogen and helium, and so being composed almost exclusively of the less abundant elements.

1.3 MEAN ORBITAL ELEMENTS OF THE PLANETS

The planets revolve about the Sun in its equatorial plane to within 4°, except for Mercury (7°) and Pluto (17.2°). Interestingly, these are the inner- and outermost planets. The orbits are closely circular, though not exactly so. The

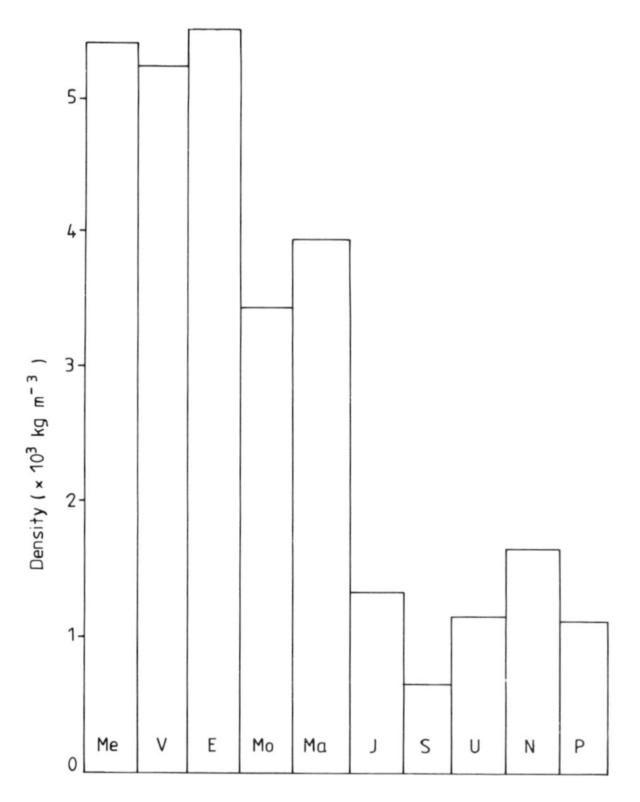

Figure 1.3 *Spread of densities for Mercury (Me), Venus (V), Earth (E), Moon (Mo), Mars (Ma), Jupiter (J), Saturn (S), Uranus (U), Neptune (N) and Pluto (P).*

mean orbital radius of the Earth about the Sun has the magnitude 1.496×10^8 km and this distance is virtually the astronomical unit (denoted by AU, being the magnitude of the semi-major axis of the Earth's orbit). The terrestrial planets have orbital radii less than 1.6 AU, while the Jovian planets have orbital radii greater than 5 AU. This distribution gains significance when viewed in conjunction with the conclusions summarised in table 1.3. The various orbital data are collected in table 1.4. Taken in conjunction with the data contained in table 1.1, table 1.4 shows the Solar System to be a very empty place indeed in terms of planetary matter. Expressed another way, the initial condensation process at the formation of the Solar System was either surprisingly complete or the material which did not condense was efficiently removed from the Solar System.

Table 1.4 *The physical elements of the planetary orbits, including mean radius, orbital period, mean orbital velocity, orbital eccentricity and inclination to the ecliptic.*

Planet	Mean radius (Earth = 1)	Orbital period (days)	Mean orbital velocity ($km\,s^{-1}$)	Eccentricity	Inclination to ecliptic (degrees)
Mercury	0.387	87.97	47.89	0.2056	7.00
Venus	0.723	224.70	35.03	0.0068	3.39
Earth	1.000	365.3	29.79	0.0167	—
Mars	1.524	686.98	24.13	0.0934	1.85
Jupiter	5.203	4332.6	13.06	0.0485	1.30
Saturn	9.539	10759.2	9.64	0.0556	2.49
Uranus	19.182	30685.4	6.81	0.0472	0.77
Neptune	30.058	60189.0	5.43	0.0086	1.77
Pluto	39.44	90465.0	4.74	0.250	17.2

1.4 ORBITAL AND ROTATIONAL ANGULAR MOMENTA

Angular momentum is associated with a rotating body, and for a planet in orbit about the Sun there are the two contributions from revolution about the Sun (orbital angular momentum) and (separately) rotation about the planetary axis (rotational angular momentum). If M_P is the mass of the planet and v_P is its velocity in a circular orbit distance R from the Sun, the magnitude of the orbital angular momentum L is given by the product $L = M_P v_P R$. The magnitudes of L for the planets are listed in table 1.5, taking data from tables 1.1 and 1.4.

The magnitude of the rotational angular momentum l for the planet regarded as a rigid body is expressed by the simple product $l = I_P w$ where I_P is the moment of inertia of the planet and w the rotational angular speed. Introducing the radius R_P of the planet, we write $w = v/R_P$ where v is the equatorial speed at the surface. The moment of inertia of a spherical body is known to be expressible in the form $I_P = \alpha_P M_P R_P^2$ where α_P is the dimensionless inertia factor for the planet. α_P is a measure of the mass distribution within the planet and is a very important parameter for describing the interior conditions. For a homogeneous sphere $\alpha_P = 0.4$, while for a point mass $\alpha_P = 0$. These limiting conditions encompass the range of possibilities in between. Quite generally, the greater the concentration of mass towards the centre the smaller the inertia factor. Combining these formulae together gives, finally,

$$l = \alpha_P M_P R_P v.$$

Data for the magnitudes of l for the different planets are collected in table 1.5. It is seen that the rotational angular momentum is negligible in comparison with

Table 1.5 *Rotational momentum and distribution of matter data for the planets of the Solar System. L is the orbital angular momentum, v the equatorial speed at the surface of the planet, l the rotational angular momentum of the planet and α_P the moment of inertia factor.*

Planet	L (kg m^2 s^{-1})	v (m s^{-1})	l	α_P
Mercury	9.15×10^{37}	—	—	—
Venus	1.85×10^{40}	—	—	—
Earth	2.66×10^{40}	4.65×10^2	5.87×10^{33}	0.331
Mars	3.53×10^{39}	2.41×10^2	1.99×10^{32}	0.376
Jupiter	1.93×10^{43}	1.27×10^4	4.59×10^{39}	0.264
Saturn	7.82×10^{42}	3.68×10^3	2.64×10^{37}	0.21
Uranus	1.69×10^{42}	2.94×10^3	1.53×10^{35}	0.23
Neptune	2.51×10^{42}	2.73×10^3	2.02×10^{35}	0.29
Pluto	3.91×10^{38}	1.611×10	—	—

the orbital contribution. The problem of deducing the distribution of mass within a planet will concern us in chapters 3 and 4 but we anticipate this work and list data for α_P in table 1.5.

For comparison, the rotational angular momentum for the Sun, l_S, is deduced from $\alpha_S = 0.05$, $M_S = 1.99 \times 10^{30}$ kg, $R_S = 6.96 \times 10^8$ m and $v_S = 1.875 \times 10^3$ m s^{-1} to be $l_S = 1.3 \times 10^{41}$ and is the only significant contribution to the solar angular momentum. (We have neglected the small angular momentum of the Sun about the centre of mass of the Solar System.) It is two orders of magnitude lower than the orbital momentum L for Jupiter alone and barely one order of magnitude greater than the orbital contribution for the Earth. We see that 99% of the angular momentum of the Solar System resides in the planets and not in the Sun, and that Jupiter makes the largest single contribution to the angular momentum content, being slightly more than half the total of 3.14×10^{43} kg m^2 s^{-1} for all the planets combined.

1.5 SATELLITE MASSES AND RADII

The various space probe missions carried out over the last decade or so have provided a wealth of data concerning the wide range of satellites encircling the planets. The satellites are distributed unevenly throughout the Solar System. There are only three associated with the terrestrial planets, namely the Moon (with the Earth) and Phobos and Deimos (with Mars). In contrast, there are at least 29 satellites circling the major planets. Two thirds of these have a radius in excess of 100 km. Various data are collected in table 1.6 and figures 1.4 and 1.5(a) for the satellites of Earth, Mars, Jupiter and Saturn, which are the objects

Table 1.6 *Data for the satellites of those planets that have been the subject of study by space probes.*

Planet	Satellite	Mean distance from planet (km)	Mean distance in terms of planetary radius	Radius (km)	Mass (kg)	Mean density (10^3 kg m^{-3})
Earth	Moon	3.84×10^5	60.09	1.738×10^3	7.35×10^{22}	3.34
Mars	Phobos	9.380×10^3	2.759	$19.2 \times 21.4 \times 27.0$	9.6×10^{15}	1.90
	Deimos	2.350×10^4	6.912	$11 \times 12 \times 15$	2.0×10^{15}	2.10
Jupiter	Io	4.126×10^5	5.74	1.816×10^3	8.916×10^{22}	3.55
	Europa	6.709×10^5	9.33	1.563×10^3	4.873×10^{22}	3.04
	Ganymede	1.070×10^6	14.88	2.638×10^3	1.490×10^{23}	1.93
	Callisto	1.880×10^6	26.15	2.410×10^3	1.064×10^{23}	1.81
Saturn	Mimas	1.850×10^5	3.08	1.95×10^2	3.76×10^{19}	1.20
	Enceladus	2.380×10^5	3.97	2.50×10^2	7.40×10^{19}	1.10
	Tethys	2.950×10^5	4.92	5.25×10^2	6.26×10^{20}	1.00
	Dione	3.770×10^5	6.28	5.60×10^2	1.05×10^{21}	1.40
	Rhea	5.270×10^5	8.78	7.65×10^2	2.28×10^{21}	1.30
	Titan	1.222×10^6	20.37	2.560×10^3	1.36×10^{23}	1.90
	Hyperion	1.481×10^6	24.68	1.50×10^2	1.10×10^{20}	0.78
	Iapetus	3.560×10^6	59.33	7.20×10^2	1.93×10^{21}	1.20

that have been studied so far at close range using space probes. The satellites of Uranus and of Neptune, unfortunately not yet subject to close scrutiny *in situ*, are described in table 1.7 and figure 1.5(*b*). These data are, by their nature, less reliable than those of table 1.6. The entry of the Moon in table 1.6 is meant to provide a link between the data for satellites and for the planets. The four Galilean satellites of Jupiter (Io, Europa, Ganymede and Callisto), Titan of Saturn, and Triton of Neptune are all at least comparable to the Moon in size and mass, and Ganymede, Callisto, and Titan are even comparable to Mercury in size. There is no doubt that these six satellites, together with the Moon, would have been regarded as ordinary planets were they to have orbited the Sun rather than the planets as they do. The densities of Io and of Europa are comparable to that of the Moon (suggesting a general silicate composition as for the Moon) but the densities of the others are much lower, suggesting at least a partial ice composition. We shall have more to say about this in chapter 9. The point to be made here is that there is no sharp distinction to be drawn between planets and satellites as far as physical properties are

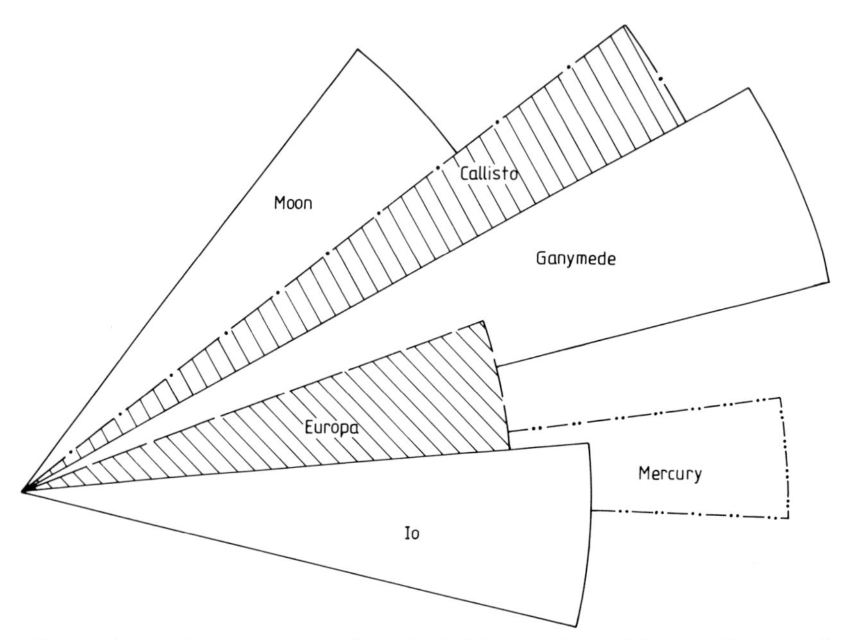

Figure 1.4 *The relative sizes, to scale, of the Galilean satellites of Jupiter. The sizes of the Moon and Mercury are included for comparison.*

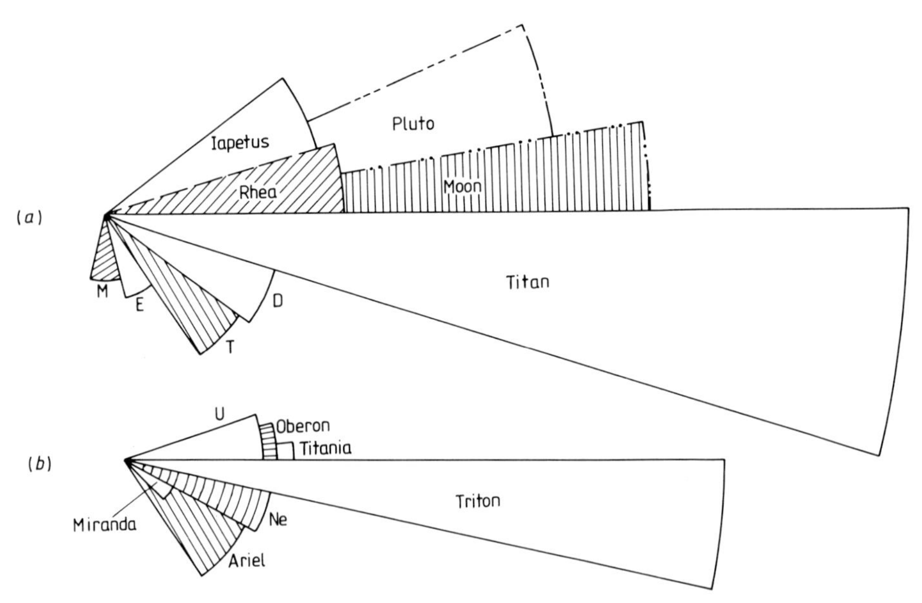

Figure 1.5 *The relative sizes, to scale, of the major satellites of* (*a*) *Saturn and* (*b*) *Uranus and Neptune. The sizes of the Moon and Pluto are included for comparison.* (*M* = *Mimas, E* = *Enceladus, T* = *Tethys, D* = *Dione, U* = *Umbriel, Ne* = *Nereid*).

Table 1.7 *Current data for the satellites of Uranus and Neptune. Improved data can be expected to result from the Voyager fly-bys of these planets later this century.*

Planet	Satellite	Mean distance from planet (km)	Radius (km)	Mass (kg, approx)	Mean density ($\times 10^3$ kg m^{-3})
Uranus	Miranda	1.30×10^5	1.60×10^2	3.4×10^{19}	
	Ariel	1.92×10^5	4.30×10^2	6.7×10^{20}	all probably
	Umbriel	2.67×10^5	4.50×10^2	7.6×10^{20}	close to 2.0
	Titania	4.38×10^5	5.2×10^2	1.2×10^{21}	
	Oberon	5.86×10^5	4.60×10^2	8.2×10^{21}	
Neptune	Triton	3.55×10^5	1.9×10^3	5.7×10^{22}	2.0
	Nereid	5.562×10^6	2.35×10^2	7.1×10^{19}	1.3

concerned. The one category merges imperceptibly into the other to form a single group of objects of varying character from the largest (Jupiter) to the smallest satellites (such as Mimas or Iapetus).

1.6 COMMENTS ON ASTEROIDS AND METEORITES

The region surrounding the Sun between the orbits of Mars and Jupiter is occupied by a belt of asteroids, although a limited number of asteroid orbits occur outside this region. Asteroids are objects with sizes ranging from a few hundred kilometres downwards. The larger members can be studied from Earth-based observatories, but the wider study would involve space probes. They are too small to be regarded as planets and so do not strictly fall in our range of discussion but some comments can be made about their composition and this is certainly of interest to us.

The masses and radii of even the largest asteroids are known only approximately. The masses of the larger members of the asteroid belt (Ceres, Pallas and Vesta) are of the order of 10^{20} kg, giving densities of 2300 kg m^{-3} (Ceres), 2600 kg m^{-3} (Pallas) and 3300 kg m^{-3} (Vesta). The magnitudes of the densities are appropriate to a silicate composition, possibly in an open (pumice) form. They are too small for pure iron, but would not be incompatible with a silicate–iron mixture. There is, indeed, evidence from the analysis of radiation from the surface at infrared wavelengths that Ceres, Pallas and Vesta are composed of a mixture of silicate and ferrous materials. The magnitudes of the masses are comparable to those of the smaller satellites of Saturn and are larger than those for the two satellites of Mars. The radii are sufficiently large ($\sim$400 km) for these three objects to be regarded as very small planetary bodies. The remaining asteroids are smaller.

Meteorites are closely related to asteroids and there is evidence that some of these bodies are debris which results from the collisions between asteroids. It can be expected that the material composition of meteorites is closely related to that from which the planets themselves were formed and the non-volatile material of the Earth and other terrestrial planets can be expected to be related directly to the meteorite composition.

The detailed classification of meteorites is a complicated process and most meteorites studied so far are unique in some particular detail. In broad terms, however, there are general classes and we can recognise meteorites which are composed exclusively of ferrous materials, called irons, and those which contain silicates as well. The minerals most commonly found in meteorites will be of interest for planetary studies and are listed in table 1.8. The remaining meteorite types involve silicates as well. A large number contain small chondrules of material about 1 cm across and are in consequence called chondrites. The average composition of a chondrite is set out in table 1.9. The second column (headed metal) refers to the free metal and is almost identical to the composition of meteor irons. The third column lists elements found in the silicate material and the final column is the average composition of the meteorite as a whole. Comparison with the entries in table 1.2 shows a close similarity in general terms, except that the volatile elements H, He, N and Ne, together with C, are absent. There are separately carbonaceous chondrites whose composition is even closer to that of table 1.2 neglecting the four gaseous elements also absent from table 1.9.

Table 1.8 *The minerals most commonly found in meteorites.*

Mineral	Composition
Kamacite	(Fe, Ni) (4–7% Ni)
Taemite	(Fe, Ni) (30–60% Ni)
Troilite	FeS
Olivine	$(Mg, Fe)_2SiO_4$
Orthopyroxine	$(Mg, Fe)SiO_3$
Pigeonite	$(Ca, Mg, Fe)SiO_3$
Diopside	$Ca(Mg, Fe)Si_2O_6$
Plagioclase	$(Na, Ca)(Al, Si)_4O_8$

Table 1.9 *Mean chemical composition for a chondrite in terms of mass (per cent).*

Element	Metal	Silicate	Average overall
O	—	43.7	33.2
Fe	90.7	9.88	27.2
Si	—	22.5	17.1
Mg	—	18.8	14.3
S	—	—	1.93
Ni	8.8	—	1.64
Ca	—	1.67	1.27
Al	—	1.60	1.22
Na	—	0.84	0.64
Cr	—	0.51	0.39
Mn	—	0.33	0.25
P	—	0.14	0.11
Co	0.48	—	0.09
K	—	0.11	0.08
Ti	—	0.08	0.06

1.7 TERRESTRIAL AND LUNAR COMPOSITIONS

Direct compositional analyses of the surface regions of the Earth can be obtained to a depth of several tens of kilometres by drilling; such an analysis for the Moon has resulted from the surface rock material brought back by the Apollo astronauts. The composition below the outer regions of the crust can be deduced only indirectly, and especially by drawing analogies with meteorite compositions. First guesses about the core composition might be deduced from iron meteorites (see the second column of table 1.9) and ideas about the deep crustal composition can be guided by the direct analysis of material collected from the continental and oceanic regions. The data regarded as broadly reliable for the crust are collected in table 1.10. Comparison between tables 1.10 and 1.2 show that on this basis the crust does not follow the cosmic abundance (which is not unexpected) and that the material is generally less dense than the mean composition for the Earth as a whole.

A likely full Earth composition is set out in table 1.11. This, it must be stressed, is no more than a guess of the overall composition, being one of a number of possibilities; these do, however, differ only in detail and agree in broad outline. The data of table 1.11 must be viewed in this general way.

The lunar composition is generally similar to that of the Earth but with some differences. For instance, the comparison of the surface basalts from the Earth and the Moon show a broadly comparable composition, but with the lunar material depleted systematically in the volatile elements. These have a lower melting point than the non-volatiles and they include low condensation

Table 1.10 *Relative abundances (silicon = 1) by mass and by number of atoms for terrestrial crustal material.*

Element	Relative abundances (Si = 1)	
	by mass	by number
O	1.61	2.82
Si	1.00	1.00
Al	2.9×10^{-1}	2.95×10^{-1}
Fe	1.67×10^{-1}	8.18×10^{-2}
Ca	1.42×10^{-1}	1×10^{-1}
K	8.3×10^{-2}	5×10^{-2}
Na	7.96×10^{-2}	9.5×10^{-2}
Mg	6.57×10^{-2}	7.3×10^{-2}
Ti	1.73×10^{-2}	9×10^{-3}

Table 1.11 *An assessment of the general composition of the Earth as regards crust, mantle and core (after W G Ernst).*

Oxide	Crust		Mantle	Core
	Continental	Oceanic		
Quartz (SiO_2)	60.1	49.9	38.3	—
Brookite (TiO_2)	1.1	1.5	0.1	—
Alumina (Al_2O_3)	15.6	17.3	2.5	—
Haematite (Fe_2O_3)	3.1	2.0	—	—
Ferrous oxide (FeO)	3.9	6.9	12.5	—
Ferrous sulphide (FeS)	—	—	5.8	—
Iron (Fe)	—	—	11.9	90.8
Nickel (Ni)	—	—	1.4	8.6
Cobalt (Co)	—	—	0.1	0.6
Magnesium oxide (MgO)	3.6	7.3	24.0	—
Lime (CaO)	5.2	11.9	2.0	—
Sodium oxide (Na_2O)	3.9	2.8	0.2	—
Potassium oxide (K_2O)	3.2	0.2	0.2	—
Phosphoric anhydride (P_2O_5)	0.3	0.2	0.2	—

elements such as bismuth, thallium, cadmium, zinc, gold, silver and lead, among others. These results are collected in table 1.12. Comparisons of this type would be particularly relevant for studies of the formation of the Solar System and of its early history.

Table 1.12 *Comparison of the relative abundances of various volatile and refractory elements in terrestrial and lunar basalts. The greater abundance of refractory elements in the Moon and volatile elements in the Earth is clear.*

Element	Condensation temperature (°C)	Ratio of abundances; Moon/Earth
Volatile elements		
Bismuth	220	11.5×10^{-3}
Thallium	195	4×10^{-3}
Calcium	400	7×10^{-3}
Zinc	530	8.5×10^{-3}
Gold	680	9×10^{-3}
Silver	495	1×10^{-2}
Rubidium	890	3.5×10^{-2}
Indium	80	3.8×10^{-2}
Potassium	890	6.5×10^{-2}
Germanium	690	6.9×10^{-2}
Sodium	890	8×10^{-2}
Lead	230	9×10^{-2}
Copper	780	1.05×10^{-1}
Gallium	590	3×10^{-1}
Manganese	950	1.8
Sulphur	400	4.4
Refractory elements		
Iridium	1350	1.1×10^{-1}
Nickel	1150	4×10^{-1}
Aluminium	1300	8×10^{-1}
Silicon	1150	9.5×10^{-1}
Calcium	1370	1.6
Iron	1370	1.9
Magnesium	1200	1.9
Titanium	1390	2
Thorium	1090	4.2
Uranium	1280	5
Barium	1100	6
Chromium	1100	10.5

1.8 MORE GENERAL OBSERVATIONS

The first direct data on the surface composition of Venus have been obtained from two Soviet soft remote landers, Venera 13 and 14. Results of the analysis

Table 1.13 *Relative abundances of oxides on the Venus surface at two sites.*

Oxide	Relative abundance (% mass)	
	Venera 13 site	Venera 14 site
SiO_2	45	49
Al_2O_3	16	18
MgO	10	8
FeO	9	9
CaO	7	10
K_2O	4	0.2
TiO_2	1.5	1.2
MnO	0.2	0.16

Table 1.14 *Relative abundances of thirteen elements normalised to hydrogen (=12.00) for the Solar System, the Orion nebula (probably representative of the interstellar medium) and an average composition for a planetary nebula. The three compositions can be taken to be the same.*

Element	Solar System	Orion Nebula	Planetary Nebula
H	12.00	12.00	12.00
He	10.9	11.04	11.23
C	8.6	8.37	8.7
N	8.0	7.63	8.1
O	8.8	8.79	8.9
F	4.6	—	4.9
Ne	7.6	7.86	7.9
Na	6.3	—	6.6
S	7.2	7.47	7.9
Cl	5.5	4.94	6.9
Ar	6.0	5.95	7.0
K	5.5	—	5.7
Ca	6.4	—	6.4

of surface rock samples by x-ray fluorescence methods were transmitted back to Earth and some results are collected in table 1.13. The overall similarity with the corresponding data of table 1.11 is truly remarkable. The Venera 14 site has a composition closely similar to that of terrestrial oceanic crust and there are distinct similarities between the Venera 13 site and the terrestrial continental crust. This first direct comparison of the composition of Venus and

Earth, often for reasons of size and density regarded as twin planets, has proved most exciting and gives support for a common composition for the terrestrial planets.

It is important to test the validity of the universal cosmic abundance of the elements associated with the Solar System as widely as possible and a comparison between the compositions of the Solar System, the nebula in Orion (M42) and a representative planetary nebula is given in table 1.14. There is a very close similarity between the compositions of these three very different objects which gives confidence in the belief in a universal cosmic abundance of the elements.

1.9 CONCLUSIONS

There are several points to notice for our further analysis.

1 The components of the Solar System show an overall unity of size, mass and composition. This unity is expressible in terms of a cosmic abundance of the elements of universal applicability.

2 The planets fall into two groups: the terrestrial planets, with masses in the range 10^{23}–10^{24} kg, and the major planets, with masses in the range 10^{26}–10^{27} kg. Pluto, the outermost planet, is a much less massive object ($\sim 10^{22}$ kg) and is anomalous in several ways.

3 The radii of the planets are more nearly similar than the masses, being in the range 10^3 to nearly 10^5 km. The radius of the Sun is about 7×10^5 km.

4 The densities of the terrestrial planets are all greater than 3000 kg m^{-3} while those for the other planets are less than 2000 kg m^{-3}.

5 The angular momentum of the Solar System, including the Sun, resides almost entirely in the major planets and particularly in Jupiter, which carries rather more than 50% of the total.

6 The angular momentum of a planet–satellite system has the different characteristic that the central planet contains the major part of the angular momentum, the satellite system containing relatively very little.

7 The planets can be regarded as a collection of nucleons (protons and neutrons) and electrons arranged in atomic combinations, and the masses can be expressed in terms of numbers of nucleons. In these terms Jupiter is a collection of some 10^{54} nucleons and this is the largest planetary accumulation. The Earth contains some 10^{51} nucleons and the smallest bodies (the largest asteroids) are composed of about 10^{47} nucleons. For comparison, the Sun contains around 10^{57} nucleons.

8 The satellites of the Solar System are involved with the major planets

except for the Moon and Phobos and Deimos (of Mars). The largest of them have masses as high as 10^{23} kg and three of them are as large as Mercury.

9 On the basis of density, Jupiter and Saturn must have a composition close to, though not identical with, the cosmic abundance. This means they are made up very largely of hydrogen and helium. The other major planets (that is Uranus and Neptune) must contain a significant proportion of heavier elements. The terrestrial planets, Pluto, the satellites and asteroids are composed primarily of the rarer elements of the cosmic abundance and are deficient in hydrogen and helium. Outside the major planets, helium is rare and hydrogen occurs very largely in combination with oxygen in the form of water.

10 The largest asteroids have masses in the range 10^{20}–10^{21} kg and radii of a few hundred kilometres. The densities can be interpreted as implying a silicate composition, at least in part.

11 The meteories have a general ferrous and silicate composition.

12 In terms of material density the Solar System is a very empty region of space. This can be taken as evidence either of the propensity of matter to condense in the very early period of the formation of the Solar System or the efficient removal of uncondensed matter.

REFERENCES AND COMMENTS

RC1.1. PLANETARY MASSES AND RADII

The physical data for the planets and satellites are contained in a wide variety of books. As examples see especially:

Beatty J K, O'Leary B and Chaikin A eds 1981 *The New Solar System* (London: Cambridge University Press)
Mitton S ed 1977 *The Cambridge Encyclopaedia of Astronomy* (London: Jonathon Cape)

These books contain a very substantial amount of information about the Solar System, and the first one much of more general import.

Other books of a general character are:

Cole G H A 1978 *The Structure of Planets* (London: Wykeham Press)
Cook A H 1980 *Interiors of the Planets* (London: Cambridge University Press)
Briggs G and Taylor F 1982 *Photographic Atlas of the Planets* (London: Cambridge University Press)

This atlas contains a range of excellent photographs of the planetary surfaces obtained from space probes together with a comprehensive survey of the field. Another photographic exploration of the planets is contained in

Guest J with Butterworth P, Murray J and O'Connell W 1979 *Planetary Geology* (London: David and Charles)

The proceedings of the twenty third COSPAR Workshop is

Stiller H and Sagdeev R Z eds 1981 *Planetary Interiors* (Oxford: Pergamon)

This is a concise and detailed account of research activities at that time.

RC1.2 THE COSMIC ABUNDANCE

This is considered by

Mitton S ed. 1977 *The Cambridge Encyclopaedia of Astronomy* (London: Jonathon Cape)

See also:

Tayler R J 1972 *The Origin of the Chemical Elements* (London: Wykeham)

for a wide and readable account. Earlier work is given by

Cameron A G W 1973 Abundances of the elements in the Solar System *Space Sci. Rev.* **15** 121–46

There is also a discussion by

Hoyle, Sir Fred 1978 *The Cosmogony of the Solar System* (Cardiff: University College Cardiff Press)

A wide ranging review of the mineralogy of the Solar System with many related topics is given in

Smith J V 1979 Mineralogy of the Planets: a voyage in space and time *Mineral. Mag.* **43** 1–89

This paper includes a comprehensive list of references.

RC1.3 MEAN ORBITAL ELEMENTS

The planets and satellites follow orbits determined by the action of the force of gravity as described by Newton's law of universal gravitation. This classic work is well presented by

French A P 1971 *Newtonian Mechanics* (New York: Norton)

The paths of the planets round the Sun are described by Kepler's laws of planetary motion, which also apply (with the appropriate modification of language) to the motion of a satellite round a planet as centre. These laws are

1 The path of a single small mass round a central, stationary, indefinitely large mass is a plane ellipse with the central mass at one of the two foci.
2 The hypothetical line joining the masses sweeps out, during the orbit, equal areas in equal times.
3 The square of the time for the small mass to orbit the larger stationary mass once is proportional to the cube of the mean distance between them (length of the semi-major axis).

It is presumed that the central sun and the orbiting planet are sufficiently small to be regarded as essentially point masses with no internal structure. These laws can be expressed alternatively in terms of conserved quantities and are the direct consequence of the following three statements: the instantaneous force between the masses is inversely proportional to the square of the distance separating them; the energy of the system is conserved (laws 1 and 3 are alternative statements of this); and the angular momentum of the system is conserved (law 2 is an expression of this).

Gravitational attraction is a universal phenomenon and there is not in the Solar System a simple Sun–planet or planet–satellite interaction. The motion of each body is perturbed to a greater or lesser extent by the other members and Kepler's laws are a generalisation not found in practice if high accuracy is applied to the description of the observed trajectories. This wider perturbation problem of celestial mechanics is very well explained in

Ryabov Y 1961 *An Elementary Survey of Celestial Mechanics* (New York: Dover)

The theory of planetary orbits has seen the most general developments in the theory of mechanics. This is one of the most beautiful constructs of the human mind and has been explained in a wide variety of books; one of the best accounts, which contains much of the excitement of the study, is

Lanczos C 1954 *Variational Principles of Mechanics* 2nd edn (Toronto: Toronto University Press)

RC1.4 ROTATIONAL ANGULAR MOMENTUM

The moment of inertia I of a mass m about an axis of rotation distance r away is defined to be $I = mr^2$. The moment of inertia of a distribution of mass about a common rotation axis is the sum of the contributions of all the masses. For a continuous distribution of mass, such as is found in a planetary body, the summation is made over the entire volume, and then we have

$$I = \int_V r^2 \, \mathrm{d}m = \int_V r^2 \rho \, \mathrm{d}V.$$

The dimensions of I are those of (mass) × (distance)^2 and the moment of inertia is expressed in dimensionless form by taking the ratio $I/M_{\mathrm{P}}R_{\mathrm{P}}^2$ for the planet as a whole. This ratio is called the inertia factor α_{P} for the planetary body.

For a further development see §3.6 and RC4.3.

RC1.5 COMPOSITION ASTEROIDS AND METEORITES

King E A 1976 *Space Geology* (New York: Wiley)

This also gives an account of the interior compositions of the Earth and Moon. For a further discussion see

Taylor S R 1975 *Lunar Science: a post-Apollo View* (New York: Pergamon)

RC1.6 RADIOACTIVE DATING

An important aspect of planetary problems is that of dating the rock material, and radioactive methods remain the most important and reliable.

If there are initially N_0 radioactive atoms in a given fixed volume of material, the number $\mathrm{d}N$ of atoms which decay during the time interval $\mathrm{d}t$ is found to be proportional to both N_0 and $\mathrm{d}t$. We can therefore write

$$\mathrm{d}N = -\lambda N_0 \, \mathrm{d}t \qquad \text{(RC1.1)}$$

where λ is a decay constant independent of both N_0 and $\mathrm{d}t$ and the negative sign is introduced to show that initial atoms are being lost. λ is characteristic of the particular radioactive atoms under decay and has the dimensions of $(\text{time})^{-1}$. It expresses the rate of decay and is large for slow decay and small for fast decay.

Radioactive decay can involve the emission of hadrons, leptons (both particles and anti-particles) and radiation in a combination which is characteristic of the particular atomic structure under decay. The hadron is the alpha-particle ($A=4$, $Z=2$) and is a helium nucleus. The leptons can be either electrons or positrons together with neutrinos. The essentially massless neutrinos carry energy and angular momentum. The energy of the hadron emission is of the order 10^6 eV and has a discrete spectrum corresponding to a quantised nucleus. The magnitude of the energy associated with the nucleus is such that the radiation lies in the gamma-ray region of the electromagnetic spectrum, this region playing for nuclear processes the same role as visible light in atomic rearrangements.

The loss of an alpha-particle reduces A by 4 and Z by 2, so we can write for the element X

$${}_A\mathrm{X}^Z \rightarrow {}_{A-4}\mathrm{X}^{Z-2} + \alpha.$$

The loss of an electron (positron) leaves A unchanged but Z is increased (decreased) by unity. Then

$${}_A\mathrm{X}^Z \rightarrow {}_A\mathrm{X}^{Z+1} + \text{electron} + \text{anti-neutrino}$$

$${}_A\mathrm{X}^Z \rightarrow {}_A\mathrm{X}^{Z-1} + \text{positron} + \text{neutrino}.$$

The emission of gamma-rays leaves A and Z unaffected, although the energy of the atom is changed. It is seen that radioactive decay involving particle emission changes the chemical nature of the atom. The initial (parent) atom decays into a daughter atom which itself may be unstable. There can be a chain of daughter atoms forming a chain of radioactive decay which stops when a stable product is reached. The uranium isotopes ${}_{92}U^{235}$ and ${}_{92}U^{238}$ are two parent elements which decay through a series of daughter nuclei to become different isotopes of lead as final stable nuclei. The energy balance in such chains, including the emitted particles and radiation, is made in energy units by using the Einstein relation $E = mc^2$, where E is the energy associated with a mass m and c is the speed of light.

For geophysical applications the elements U^{235}, U^{238}, K^{40} and Th^{232} are especially relevant to the long-term energy budget, having decay times in excess of 10^8 years. Other elements can be of interest in particular circumstances.

For an extended time period T, the integration of equation (RC1.1) gives

$$N = N_0 \exp(-\lambda T) \qquad \text{(RC1.2)}$$

for the number N of atoms remaining to decay. For a given material (so that λ is given), the time T during which the particular decay process has been taking place unhindered can be deduced if N_0 and N are known. Even if N (the daughter product) can be measured, N_0 (the parent) remains unknown. Information about the initial parent concentration can be deduced if none of the daughter product has been able to escape from the volume during the period T and so the number DN of atoms that have decayed in the sample during the time T can be found. Then

$$N_0 = N + DN = N \exp(+\lambda T)$$

using equation (RC1.2). With N, λ and DN known, T can be deduced. Explicitly

$$T = (1/\lambda) \ln(1 + DN/N).$$

The logarithm in this expression can be simplified if $DN \ll N$ since then $\ln(1 + DN/N) \approx DN/N$. This will apply, for instance, when the decay rate is very slow. A useful way of specifying the decay rate is through the half-life. This is the time $t_{1/2}$ taken for half the radioactive atoms in an initial sample to decay. We integrate equation (RC1.2) over this time with the condition that $N = N_0/2$. The result is

$$t_{1/2} = \ln 2/\lambda.$$

The daughter product may itself be radioactive and this adds a complication in practice, although the same arguments can be applied to the daughter as to the parent material. Again, different isotopes of the same element will behave differently under radioactive decay and a study of the isotopes ratios of different materials will also be a powerful way of approaching this problem. The radioactive decay method will, of course, be a reliable method of estimating time scales only if the daughter products are trapped firmly in the volume under test. This may be a rock sample but it may be a more extended volume, for instance the atmosphere of a planet itself.

This whole approach has been developed into an important branch of geophysics over the last two decades and has particularly exploited the accuracy of mass spectrometric techniques. The reader is referred to the specialised literature for details and a full account: two starting sources are

Garland G D 1979 *Introduction to Geophysics* 2nd edn (Philadelphia: Sanders)

which contains a good discussion and many references; and

Glass B P 1982 *Introduction to Planetary Geology* (London: Cambridge University Press) pp 37–41

A very readable and broadly popular account is given by

Ozima M 1981 *The Earth: its birth and growth* (London: Cambridge University Press)

2

THE PLANETARY BODY

The system of planets and satellites is to be regarded, from a physical point of view, as a single range of objects controlled by the same physical processes. The differences between them are interpreted as a manifestation of the consequences of scale and detailed composition. The fact that some objects orbit the Sun while others orbit individual planets is immaterial to the physical nature of their interiors. In order to express the general nature of these objects it is convenient to introduce the terminology of planetary body rather than use the traditional division of planet and satellite.

2.1 PLANETARY BODY

When enough matter is brought together, the mutual gravitational attraction between its component parts is greater than the opposing tendency of the motion of the parts to separate. The critical mass necessary for a volume of gas to condense is the Jeans mass (see RC2.1) and a comparable circumstance exists if the material is in the form of microscopic particles rather than gas. The effect of the gravitational forces is to compress the material into as small a volume as possible. The magnitude of the minimum size will be obtained when the constituent atoms and molecules are compressed together. The degree of compression is determined by the strength of the gravitational force which itself increases with the mass of material. The equilibrium structure is in this way to be expressed as a balance between the gravitational forces pulling the material together and the resistance of the material to further compression. This resistance is expressed in terms of an incompressibility which is ultimately a property of the individual atoms although it can be expressed macroscopically in terms of a bulk modulus of elasticity. A body whose equilibrium size is determined by the balance between the gravitational force of compression and the atomic force of incompression will be called a planetary body.

The planetary body has several characteristics and three should be noted from the beginning. First, we make no mention of thermal forces in defining the equilibrium and in this sense a planetary body is a cold body. This does not mean that its temperature is either very low or uniform throughout, but that the equation of state of the material is independent of temperature. We shall find later, in fact, that the temperature increases with depth and that the central temperature is usually several thousands or even tens of thousands of degrees for the larger members of the Solar System. The associated gradient of temperature will give rise to a heat flow through the body which may cause various material motions, such as convection. These can have important consequences for the detailed structure or evolution of the body, but play no essential part in determining its overall equilibrium size. This means that the distribution of temperature can be disregarded for many analyses of planetary interiors and the body is cold in this sense.

Second, the pressure inside a body of sufficient size will cause the atoms to break up, and the equilibrium will then involve the nucleons directly rather than collectively, as in atoms. This will be the circumstance, for instance, in a white dwarf star and in a neutron star (see RC2.4) and again the thermal energy will not contribute to the equilibrium size. Such a body will also be treated as cold. A planetary body, therefore, is a special case of a more general classification of a cold body.

Third, magnetic phenomena abound throughout the Solar System and the main geomagnetic field has been used for navigation for hundreds of years. Magnetic energy associated with a planetary body will always be small in comparison with the other energies controlling it (for instance the gravitational energy) and will make no contribution to the overall equilibrium. The planetary body is then to be treated as intrinsically non-magnetic. Even so, the study of the origin of the planetary magnetic field is of interest and importance (quite separately from its intrinsic fascination as a physical phenomenon to be understood) in providing a means of inferring conditions in the interior. Very special conditions are necessary if a magnetic field is to be produced and maintained in a planetary volume and the observation of a field in practice implies that such conditions prevail somewhere inside. These matters are the subject of chapter 6.

2.2 COMPRESSION AND THE BULK MODULUS

The equilibrium between the gravitational force of compression and the incompression of the material which characterises a planetary body can be maintained only if neither force is dominant. This provides a criterion for specifying a range of size for such a body.

2.2.1 *COMMENTS ON THE EFFECTS OF COMPRESSION*

As mass is accumulated, the gravitational energy will increase and so also will the degree of compression inside. The resistance to compression is provided by the strength of the atoms, which is a characteristic of the particular material and not of the amount there. If too little material is present, the resulting compression will be too small to affect the atomic equilibrium, while if the compression exceeds the atomic strength, atoms are broken up. In either case, the type of equilibrium between internal forces that we have described is not possible and, although the body may well be cold, we would not regard it as planetary. There is a lower limit and an upper limit to the size of a planetary body. These limits are fixed in particular cases ultimately by the properties of the atoms comprising the material. We deduce the magnitudes of these limits in §2.2.2.

Smaller bodies than the lower limit will be called planetesimals and the very smallest, dust. Bodies larger than the upper limit, where the atom has collapsed to form free nuclei and electrons, will generally be stellar objects. Thermonuclear processes can now come into play and the effect of temperature cannot be neglected—the body is no longer cold. Thermonuclear burning will stop when the fuel (hydrogen or the lighter elements) has been used up. Then a new equilibrium will be achieved, different from that found in planets but still not dependent on the temperature. The body (a white dwarf or a neutron star) is then cold once again, although the mass will be larger than that of a planetary body. In this way, conditions are specified primarily by the features of compression, or resistance to it, and for a planetary body this can be described by the ordinary bulk modulus of elastic incompression.

2.2.2 *ROLE OF THE BULK MODULUS*

At each level within the planetary body, the material there supports the weight of the material above. As a consequence, the volume taken up by a given mass of material decreases with the depth z, so the density ρ increases with depth. The density depends only on the pressure p and not on the temperature (remember we are dealing with a cold body) so we can write $\rho=\rho(p)$. This is an expression of the so-called equation of state of planetary material. For the density at the depth $z+\mathrm{d}z$ close to z we can write the expansion

$$\rho(z+\mathrm{d}z)=\rho(z)+\left.\frac{\partial\rho}{\partial p}\right|_z p(z)+\frac{1}{2}\frac{\partial}{\partial p}\left(\left.\frac{\partial\rho}{\partial p}\right|_z\right)p^2(z)+\ldots \tag{2.1}$$

allowing the difference $\rho(z+\mathrm{d}z)-\rho(z)$ to be calculated if the quantity $(\partial\rho/\partial p)_z$ is known together with its dependence on pressure.

The bulk modulus K for the pressure pertaining at the depth z is defined by

$$K(z)=\rho(z)\frac{\partial p(z)}{\partial\rho(z)} \tag{2.2}$$

being the inverse of the compressibility. Then equation (2.1) becomes

$$\rho(z+\mathrm{d}z)-\rho(z)=\frac{\rho(z)}{K(z)}\,p(z)+\frac{1}{2}\frac{\partial}{\partial p}\left(\frac{\rho(z)}{K(z)}\right)_z p^2(z)+\ldots.$$

For small differences of depth we have approximately

$$\frac{\rho(z+\mathrm{d}z)-\rho(z)}{\rho(z)}=\frac{p(z)}{K(z)}. \tag{2.3}$$

We see that K has the same dimensions as pressure, and determines the change of density with pressure. The change of density consequent upon a given increase of pressure is larger the smaller the K (i.e. the larger the compressibility). Using equation (2.2), equation (2.1) can be rewritten in the form

$$\frac{\rho(z+\mathrm{d}z)-\rho(z)}{\rho(z)}=\frac{p(z)}{K(z)}+\frac{1}{2}\frac{p^2(z)}{K^2(z)}\left(1+\frac{\partial K(z)}{\partial p}\right)+\ldots. \tag{2.4}$$

This is an expansion in powers of the pressure, but with the coefficients of the expansion determined by K and its derivatives with respect to pressure.

The expansion (2.4) can be inverted to form a series expansion for pressure in terms of density or (because we deal with a cold body) can be converted into an expansion of bulk modulus in terms of pressure. We shall use this possibility in chapter 3 as a means of generating an equation of state, giving pressure explicitly in terms of density. It will be realised from what has been said so far that a knowledge of the bulk modulus of the material within a planetary body is crucial to describing the physical conditions there.

2.3 ATOMIC COMPOSITION

An atom consists of a central positively charged nucleus of linear dimension about 10^{-15} m surrounded by an organised cloud of negatively charged extranuclear electrons extending out to some 10^{-10} m. There are A nucleons and Z positive charges in the nucleus and Z electrons outside, the atom having a mass Am_p where m_p is the mass of each nucleon, and zero net electric charge overall. The extranuclear electrons are in motion within the atom relative to the nucleus and provide a pressure which keeps the atom distended as a three-dimensional object.

2.3.1 ATOMIC SIZE

For simplicity, we consider hydrogen first because it is the simplest atom, being composed of a single nuclear proton and one extranuclear electron. If the electron of mass m_e has momentum $\boldsymbol{p}$ relative to the nucleus, the energy E of the atom is

$$E = \frac{p^2}{2m_e} - \frac{e^2}{4\pi\varepsilon_0 r} \tag{2.5}$$

where r is the distance between the nucleus and some appropriately chosen location for the electron. The factor $4\pi\varepsilon_0$ is the permittivity appropriate to SI units $(=\frac{1}{9} \times 10^{10})$ and the unit electronic charge e is measured in coulombs. The second term in equation (2.5) is the potential energy of attraction between the two charges $+e$ and $-e$ and, being an attraction, carries a negative sign.

The electron within the atom is not a simple particle but fills the atomic volume, supposed to be a sphere of radius R. The rapid motion of the electron is periodic and is associated with a wave of wavelength λ according to the relationship of de Broglie

$$p\lambda = h, \tag{2.6}$$

h being Planck's constant. Because the electron cannot pass through the nucleus, the maximum value of λ is determined by the circumference of the atom which is proportional to R. The maximum value of λ is associated with a minimum magnitude for the momentum (according to equation (2.6)) and so to a minimum magnitude for the energy E of the atom. From these arguments the expression for the energy (equation (2.5)) is written alternatively in terms of the atomic radius

$$E \sim \frac{h^2}{8\pi^2 m_e} \frac{1}{R^2} - \frac{e^2}{4\pi\varepsilon_0 R} \tag{2.7}$$

where R is the atomic (mean) radius.

The stability of the atom requires that E should be a stationary value, by which we mean that R should be such that

$$\frac{\mathrm{d}E}{\mathrm{d}R} = 0. \tag{2.8}$$

Using equation (2.7) this value of R, denoted by a_0 and called the Bohr radius, is found to be

$$a_0 = \frac{h^2 \varepsilon_0}{\pi m_e e^2}. \tag{2.9}$$

2.3.2 *ENERGY OF THE ATOM*

The corresponding minimum energy E_0 is obtained by inserting equation (2.9) into equation (2.7) in place of R: the result is

$$E_0 \sim -\frac{1}{8} \frac{m_e e^4}{h^2 \varepsilon_0}. \tag{2.10}$$

The negative sign refers to an energy of attraction, which is necessary for the atom to be bound overall as a stable structure. The particular energy E_0 is called the Rydberg energy and is often denoted by Ryd. It is a measure of the energy associated with the atom. Inserting values for the various constants in equation (2.10) we find $|\mathrm{Ryd}| \sim 10^{-18}$ J. The size a_0 and energy Ryd are the two important characteristics of the atom from the present point of view.

The same arguments will apply to atoms heavier than hydrogen. If the nucleus has a charge Ze and there are Z extranuclear electrons, the force between the nucleus and the electrons is generally increased and the atomic size is consequently affected. The corresponding dimension to the Bohr radius is

$$a_0 = \frac{h^2 \varepsilon_0}{\pi m_e Z e^2} \tag{2.11a}$$

and the associated energy is

$$\mathrm{Ryd} \sim -\frac{1}{8}\frac{m_e Z e^4}{h^2 \varepsilon_0^2}. \tag{2.11b}$$

2.3.3 *ATOMIC INTERNAL PRESSURE*

An internal electronic pressure p_a, which opposes the electrical attraction between the charges and keeps the atom extended at all times, also causes the atom to resist any decrease of size below the equilibrium stable value. This internal pressure is calculated from the dependence of energy on atomic volume according to the formula of statistical mechanics

$$p_a = -\frac{\partial E}{\partial V} = -\frac{1}{4\pi R^2}\frac{\partial E}{\partial R}. \tag{2.12}$$

E is given by equation (2.7), and the expression for the pressure is the difference of two terms:

$$p_a = p_k - p_v \tag{2.13}$$

where

$$p_k = \frac{h^2}{16\pi^3 m_e}\frac{1}{R^5} \tag{2.13a}$$

and

$$p_v = \frac{e^2}{16\pi^2 \varepsilon_0}\frac{1}{R^4}. \tag{2.13b}$$

Here p_k is a kinetic pressure which arises from the motion of the electrons and, being positive, acts to increase the atomic size; p_v is a potential contribution which arises from the electrostatic attraction and acts to reduce the atomic size. For small R, $p_k > p_v$ and atomic rearrangement is towards expansion, while for large R, $p_k < p_v$ and the atom tends to contract. The balance between the two effects, when the atom tends neither to expand nor to contract, occurs at the Bohr radius, when $R = a_0$. It is interesting that the magnitude of each component of the pressure then is $p_k = p_v \sim 3 \times 10^{12}$ N m^{-2}, which is about 30 million atmospheres. This is an astonishingly high pressure in relation to the small volume of the atom but such a strength is needed to maintain the largest planets of the Solar System.

2.4 MAXIMUM AND MINIMUM SIZE

The atomic stability will be upset as the external pressure becomes sufficiently high to neutralise the effect of the kinetic pressure. Once the pressure external to the atom has a value of the general order 10^{13}–10^{14} N m^{-2}, the atoms there are incapable of maintaining the equilibrium which defines a planetary body. The pressure at the centre is due to the weight of material above so there is here a criterion to provide a maximum planetary size allowable on the basis of the atomic properties themselves.

With regard to orders of magnitude, the pressure p_c at the centre of a planet of radius R_P is given by

$$p_c \sim g\bar{\rho}R_P \tag{2.14}$$

where g is the surface acceleration of gravity and $\bar{\rho}$ is the mean density. In terms of the mass of the planet M_P

$$g = \frac{GM_P}{R_P^2} = \frac{GM_P}{\frac{4}{3}\pi R_P^3}\frac{4}{3}\pi R_P^3$$

$$= G\bar{\rho}\tfrac{4}{3}\pi R_P^3$$

where G is the universal constant of gravitation, so that

$$p_c \sim \tfrac{4}{3}\pi G\bar{\rho}^2 R_P^2. \tag{2.15}$$

Equation (2.15) allows the magnitude of the radius to be inferred to provide a prescribed central pressure.

The cosmic abundance of the elements shows a mixture of hydrogen and helium to be the most numerous material and we might suppose the largest planet to have overwhelmingly this composition. Then, setting $\bar{\rho}_c = 1 \times 10^3$ kg m^{-3} and assuming for maximum planetary size $p_c = 1 \times 10^{13}$ N m^{-2}, equation (2.15) leads immediately to the conclusion that the maximum planetary radius will be about $R_P \sim 10^8$ m. This is very little larger than Jupiter. As a comparison, the Sun has a radius about ten times greater, with a central pressure of the general magnitude 10^{15} N m^{-2}. At this pressure, the atomic stability will certainly be lost and it is known that the solar interior can be well represented as two co-existing gases, one composed of protons and the other of electrons. It seems from these approximate arguments that Jupiter is close to being the largest planetary body that can be supported by the atomic structure of the material. We shall make the arguments more precise in §2.7.

The expression (2.15) for the pressure can provide a criterion for giving a lower limit to the size of a planetary body. For there to be any semblance of equilibrium between gravitational and elastic forces, the compaction of the atoms must be such as to call elastic forces into play, at least weakly. This will involve the compression of the atoms to some extent, but there is no sharp separation distance where this begins. Because of the features of interatomic forces, we can suppose a separation of about four atomic radii as the distance where, in general, some form of repulsion can be expected to become recognisable. From equation (2.13a) this implies an internal kinetic pressure of about 10^9 N m^{-2}. At about this pressure, the material will be held together by gravity in a condensed form but the interparticle forces will not be strong enough to maintain a solid lattice. The atoms will move relative to each other, the material showing essentially fluid behaviour. Increasing the pressure will increase the fluidity, although the associated viscous dissipation may remain very high by ordinary laboratory standards. The collection of material will assume a shape of simple symmetry, which will be a sphere if the body is not rotating but otherwise will be a shape of rotational symmetry (spheroidal). In these terms, it is convenient to define the lower limit of size of a planetary body as that body which will assume spherical symmetry under the action of gravity, given enough time. This will imply an internal pressure in excess of about 10^9 N m^{-2}, and equation (2.15) then provides an estimate of the radius of about 100 km as the smallest object of planetary material that can form a spherical shape. Objects of smaller linear dimension will have an irregular shape in general, and we will call them planetesimals. We shall consider the implications of the inherent fluidity of the material in chapter 3.

In summary, we shall suppose a planetary body to have a radius in the range 10^5 to 10^8 m, the upper limit being firm but the lower limit being to some extent arbitrary.

2.5 VARIOUS ENERGIES

2.5.1 ENERGY OF AN ATOM

The energy of the atom is the Rydberg, defined through equation (2.10). Substituting the various numerical magnitudes we find Ryd $= 2.19 \times 10^{-18}$ J, which is approximately 13.6 eV. This is the complete energy for the hydrogen atom. For an atom with Z extranuclear electrons the energy is Z times as great, or about 13.5Z eV. The Rydberg is the natural atomic unit of energy and other energies can be expressed in terms of it.

The existence of condensed phases of matter is a consequence of the attractive forces between atoms. The linking of atoms is associated with a binding energy E_b which varies from one material to another. As a generalisation, we can set for simple inorganic materials $E_b \sim 10^{-2}$ Ryd, although the numerical factor will vary by a factor of two or so from one material to another.

By making the association $E = kT$, where k is the Boltzmann constant, an energy can be linked to a temperature. The temperature corresponding to the Rydberg is then found from

$$2.19 \times 10^{-18} = kT$$

which gives $T \sim 1.6 \times 10^5$ K. We could interpret this result by presuming that the ionisation of a hydrogen atom by collisions would be a common phenomenon at temperatures above 2×10^5 K. From the point of view of planetary interiors, this temperature would correspond in its actions to a pressure of about 3×10^{12} N m^{-2} (see §2.3.3) which is outside the range to be associated with a planetary body. We can, on this basis, put an upper limit to the internal temperature of such a body, because it must be less than 2×10^5 K if the atoms are not to be broken up. It may, perhaps, be smaller by an order of magnitude.

The energy for melting and vaporisation can be estimated in analogous ways. The heat of fusion H is a measure of the energy associated with the rigidity of the material and so of the energy associated with shearing the atomic arrangement at (essentially) constant volume. For many materials of geophysical interest $H \sim 3 \times 10^5$ J kg^{-1}. Because there will be about 10^{26} atoms per kilogram, we can associate an energy of fusion of about 3×10^{-21} J with each particle, which is about 10^{-3} Ryd. This is the energy that must be supplied by gravitation in order that the material will flow easily, although an energy about one order of magnitude less will be sufficient to achieve this over a longer time scale (the effective viscosity will be high). The essential property of fluidity will therefore be associated with a binding energy per particle of about 10^{-4} Ryd which will correspond to a pressure within the body of about 10^8 N m^{-2}. This order of pressure will occur in the Earth at a depth of about 30 km. It would also be found at the centre of a body of silicate composition with a radius of about 100 km. These arguments support those of §2.4, and quantify the requirement that the planetary body shall be a figure of symmetry.

2.5.2 DEGENERACY ENERGY

A planetary body of mass M_P is composed of N_P atoms where $N_P = M_P/Am_p$ and A is the mass number. If each atom has Z extranuclear electrons there will be $N_P Z$ electrons in the body. Electrons are Fermi particles obeying the Pauli exclusion principle. This means that each one will occupy a quantum mechanical state alone, and it is this feature which gives the range of elements the varied physical and chemical properties portrayed in the periodic table of the elements. This particular electronic feature is at the root of the equilibrium of a planetary body, because the electrons provide a considerable force to remain independent and each acts to reserve a region of space to itself. Each electron can be regarded as occupying a cell of linear dimension d and we can suppose the cell to be spherical.

The energy of the electron inside an atom is independent of the temperature and so is the condition of equilibrium of a planetary body. The effect of temperature on the energy of the electrons can, therefore, be disregarded whether inside the atom or, more generally, inside the planetary body. We can, in consequence, quite generally disregard the effect of the temperature on the energy of the electrons whether they are held within the atom or whether the electron is moving generally in the body. The condition that the energy of the electron is independent of the temperature is said to be the condition of degeneracy (since it applies to any temperature). The electron motion in a planetary body can, therefore, be regarded as degenerate in this sense. The energies involved in a planetary body are sufficiently low for non-relativistic theory to apply.

The kinetic energy E_k of a single non-relativistic electron is given by

$$E_k = p^2/2m_e \tag{2.16}$$

where m_e is the electron mass. The de Broglie relation is $p\lambda = h$ (see §2.3.1) and the magnitude of λ will be determined by the volume V_c of the cell available to each electron. To the present approximation we set $\lambda = 2\pi d$ as the maximum wavelength. The corresponding minimum energy for each electron in each cell, $\mathscr{E}_c$, is then written

$$\mathscr{E}_c = \frac{h^2}{2m_e} \frac{1}{4\pi^2 d^2}. \tag{2.17}$$

The total volume V_P of the planetary body contains ZN_P electrons so the linear dimension per electron is

$$d = (Am_p/M_P)^{1/3} R_P. \tag{2.18}$$

Combining equations (2.17) and (2.18) the dimension of the electron cell is eliminated to obtain

$$\mathscr{E}_c = \frac{h^2}{8\pi^2 m_e} \left(\frac{ZM_P}{Am_p}\right)^{2/3} \frac{1}{R_P^2}. \tag{2.19}$$

There are Z electrons per atom and N_P atoms so the total degeneracy (kinetic) energy of the electrons is $E_k = ZN_P\mathscr{E}_c$ which, using equation (2.19), is easily arranged into the form

$$E_k = \left(\frac{h^2}{8\pi^2 m_e m_p^{5/3}}\right) \frac{M_P^{5/3} Z^{5/3}}{A^{5/3} R^2}. \tag{2.20}$$

The physical constants can be expressed collectively in numerical terms by introducing the quantity γ_k defined by

$$\gamma_k = \frac{h^2}{8\pi^2 m_e m_p^{5/3}} \tag{2.20a}$$

so that

$$E_k = 2.592 \times 10^6 \frac{M_P^{5/3} Z^{5/3}}{A^{5/3} R^2}. \tag{2.20b}$$

This expression shows how the degeneracy energy depends on the material composition and the amount of material present in the planetary body.

2.5.3 *ELECTROSTATIC ENERGY*

The calculation of the electrostatic energy for the nuclei and electrons comprising the planet is generally a complicated task but we can get the answer with very closely the correct numerical factor in a simple way. To do this, we consider the electrostatic energy $\mathscr{E}_e$ for each cell and then add the contributions of the cells together to obtain the total electrostatic energy E_e.

As to order of magnitude, the electrostatic energy for each cell is

$$\begin{aligned}\mathscr{E}_e &\sim -\frac{1}{4\pi\varepsilon_0}\frac{Ze^2}{d}\\ &= -\frac{Ze^2}{4\pi\varepsilon_0}\left(\frac{M_P}{Am_p}\right)^{1/3}\frac{1}{R_P}.\end{aligned}$$

A more detailed analysis gives the numerical multiplier to be 0.90 but this is close enough to unity for the difference to be neglected. The total energy is then

$$E_e = \mathscr{E}_e N_P Z$$

or, introducing the various parameters explicitly,

$$E_e = -\frac{Z^2 e^2}{4\pi\varepsilon_0}\left(\frac{M_P}{Am_p}\right)^{2/3}\frac{1}{R_P}\left(\frac{M_P}{Am_p}\right).$$

We write this as

$$E_e = -\gamma_e \frac{M_P^{4/3} Z^2}{A^{4/3} R_P} \tag{2.21a}$$

with

$$\gamma_e = \frac{e^2}{4\pi\varepsilon_0 m_p^{4/3}}. \tag{2.21b}$$

Inserting numerical values into (2.21b) gives $\gamma_e = 1.16 \times 10^8$.

2.5.4 GRAVITATION ENERGY

The gravitational energy E_g is easily calculated. For a planetary mass M_P of radius R_P it is

$$E_g = -\gamma_g \frac{GM_P^2}{R_P} \tag{2.22}$$

where γ_g is a numerical factor: explicitly $\gamma_g = (1 - \alpha_P)$ where α_P is the inertia factor introduced in §1.4. For a homogeneous sphere ($\alpha_P = 0.4$), $\gamma_g = 0.6$; for a terrestrial type planet ($\alpha_P = \frac{1}{3}$), $\gamma_g = \frac{2}{3}$; for a Jovian planet ($\alpha_P = \frac{1}{5}$), $\gamma_g = 0.8$. It is sufficient now to set $\gamma_g = 1$.

2.6 MINIMUM SIZE FOR A GIVEN MASS

We can make a preliminary assessment of our arguments by making estimates of the sizes of the planets on the basis of a simple comparison between gravitational energy and the compressional energy for each planet, comparing the calculated size with observation. We would expect the comparison to be closest for the large planets, where the degree of compression is highest, but nowhere would we expect our calculated radius to exceed the observed radius.

Let us start with the kinetic pressure (equation (2.13a)) with R chosen to be the atomic radius. This pressure must balance that due to gravitational contraction. The energy e_a associated with this pressure is

$$e_a \sim p_a V_a$$

where V_a $(= \frac{4}{3}\pi r_a^3)$ is the atomic volume and r_a the atomic radius. We set $r_a = \xi a_0$ where a_0 is the Bohr radius and ξ is a numerical factor which depends on the particular material. For the terrestrial planets we can take $\xi \sim 1.5$, while for the Jovian planets we take $\xi \sim 1$, the difference reflecting differences in the strengths of the interatomic forces between hydrogenic and silicate materials.

The number of atoms in the planet $N_P = (R_P/r_a)^3$ so the gravitational energy e_g per atom is

$$e_g \sim \frac{GM_P^2}{R_P} \frac{1}{N_P} = \frac{GM_P^2}{R_P^4} r_a^3.$$

For an equilibrium balance between gravity and the atomic internal pressure we have the condition

$$p_a V_a \sim \frac{GM_P^2}{R_P^4} r_a^3.$$

The pressure p_k is given by equation (2.13a), so

$$\frac{h^2}{16\pi^3 m_e r_a^5}\tfrac{4}{3}\pi r_a^3 \sim \frac{GM_P^2}{R_P^4} r_a^3$$

and this gives immediately the expression for R_P in the form

$$R_P^4 \sim \frac{12\pi^2 m_e G r_a^5}{h^2} M_P^2 = 1.64\times 10^{28} r_0^5 M_P^2. \quad (2.23)$$

The calculated magnitudes of R_P for the terrestrial planets are collected in table 2.1, while those for the major planets are collected in table 2.2.

Table 2.1 *Minimum radii for the terrestrial planets, calculated from equation (2.23), compared with observed data.*

Body	R_P (calculated) (km)	R_P (observed) (km)	R_P (observed) / R_P (calculated)
Mercury	1.11×10^3	2.44×10^3	2.2
Venus	4.26×10^3	6.05×10^3	1.4
Earth	4.6×10^3	6.39×10^3	1.4
Moon	5.23×10^2	1.74×10^3	3.3
Mars	1.54×10^3	3.40×10^3	2.2

Table 2.2 *Minimum radii for the major planets, calculated from equation (2.23), compared with observed data.*

Body	R_P (calculated) (km)	R_P (observed) (km)	R_P (observed) / R_P (calculated)
Jupiter	6×10^4	7.19×10^4	1.2
Saturn	5×10^4	6.0×10^4	1.2
Uranus	1.2×10^4	2.6×10^4	2.2
Neptune	1.9×10^4	2.5×10^4	1.3

It is seen from the tables that the calculated magnitudes of the radii are of the same orders of magnitude as those observed, but slightly smaller in each case. This comparison lends confidence to the overall validity of the approach.

2.7 RELATION BETWEEN MASS AND RADIUS

A reliable relationship between mass and radius is obtained by considering the balance between the three main forces acting on the planetary body; gravitational and electrostatic on the one side and the internal atomic force on the other. We can approach this balance from the point of view of the energies involved and, because both the gravitational and electrostatic forces involve the inverse square of the separation distance, we can invoke the virial theorem.

The potential energy contribution to the planet will be the sum of E_g and E_e while the kinetic energy contribution will be the degeneracy energy E_k: according to the virial theorem we write

$$E_k = \tfrac{1}{2}(E_e + E_g). \tag{2.24}$$

Using equations (2.20a), (2.21) and (2.22) this relationship takes the explicit form

$$\gamma_k \frac{M_P^{5/3} Z^{5/3}}{A^{5/3} R_P^2} \sim \tfrac{1}{2}\gamma_e \frac{M_P^{4/3} Z^2}{A^{4/3} R_P} + \tfrac{1}{2}\gamma_g \frac{G M_P^2}{R_P}.$$

Multiply this expression throughout by $(2R_P^2/M^{4/3})$ to obtain

$$2\gamma_k \frac{M_P^{1/3} Z^{5/3}}{A^{5/3}} \sim R_P\left(\frac{\gamma_e Z^2}{A^{4/3}} + \gamma_G M_P^{2/3}\right) \tag{2.25}$$

where $\gamma_G = G\gamma_g$. This provides an expression for the radius R_P once the mass M_P and the composition (Z, A) are assigned (see figure 2.1).

2.7.1 *MAXIMUM RADIUS*

The expression (2.25) allows us to draw the conclusion that the radius has a maximum value for any mass. To find this radius, we invoke the condition

$$\frac{dR_P}{dM_P} = 0, \tag{2.26}$$

well known from the calculus of maxima and minima. From the expression (2.25) we form

$$\frac{1}{R_P} \sim \frac{\gamma_e Z^2/A^{4/3} + \gamma_G M_P^{2/3}}{(2\gamma_k Z^{5/3}/A^{5/3}) M_P^{1/3}} \sim 1.29 \times 10^{-16} M^{1/3} \left(\frac{A}{Z}\right)^{5/3} + 2.24 \times 10 (ZA)^{1/3} M^{-1/3} \tag{2.27}$$

so that

$$-\frac{1}{R_P^2}\frac{dR_P}{dM_P} \sim -\frac{1}{3}\frac{\gamma_e Z^{1/3} A^{1/3}}{2\gamma_k}\frac{1}{M_P^{4/3}} + \frac{1}{3}\frac{\gamma_G A^{5/3}}{\gamma_k Z^{5/3}}\frac{1}{M_P^{2/3}}. \tag{2.28}$$

The condition (2.26) implies that the right-hand side of equation (2.28) should vanish. The corresponding expression M_c for the mass is

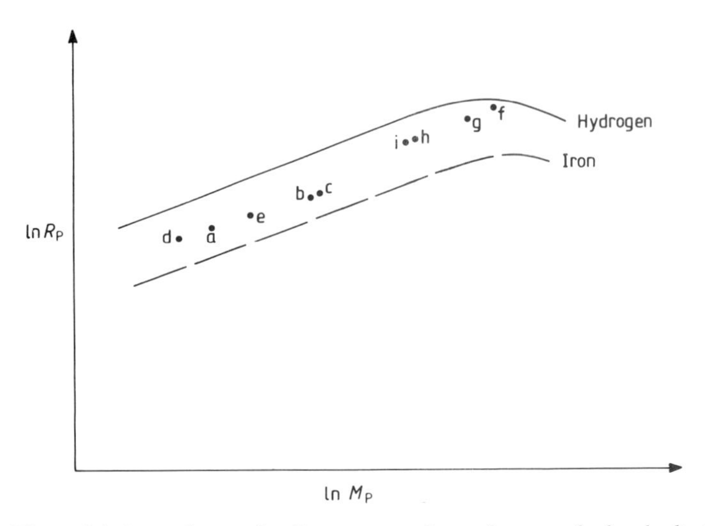

Figure 2.1 *Dependence of radius on mass for a planetary body, the limits marking a pure hydrogen and a pure iron composition. The points mark the locations for the planets on the diagram: a = Mercury, b = Venus, c = Earth, d = Moon, e = Mars, f = Jupiter, g = Saturn, h = Uranus, i = Neptune. Note that Jupiter appears close to the hydrogen line and near to the maximum permitted radius.*

$$M_c \sim \left(\frac{\gamma_e}{\gamma_G}\right)^{3/2} \frac{Z^3}{A^2} = 2.3 \times 10^{27} \frac{Z^3}{A^2} \text{ (kg).} \tag{2.29}$$

For a hydrogen sphere ($A = 1$, $Z = 1$) this gives $M_c = 2.3 \times 10^{27}$ kg, which is slightly larger than the mass of Jupiter. For silicon ($A = 28.1$, $Z = 14$) the maximum mass is 8×10^{27} kg. For an iron sphere ($A = 55.8$, $Z = 26$) the mass is $M_c \sim 1.3 \times 10^{28}$ kg.

From equation (2.27) we deduce the maximum radius to be

$$R_P(\max) \sim \frac{\gamma_k}{(\gamma_e \gamma_G)^{1/2}} \frac{Z^{2/3}}{A}$$

or in numerical terms

$$R_P(\max) \sim 1.12 \times 10^8 (Z^{2/3}/A) \text{m.} \tag{2.30}$$

For hydrogen, $R_P = 1.12 \times 10^8$ m which is slightly larger than the observed radius of Jupiter; for a pure silicon sphere the maximum radius is 2.31×10^7 m while for pure iron it is 1.76×10^7 m.

2.7.2 *MAXIMUM MASS FOR A PLANETARY BODY*

The result (2.29) used in conjunction with equation (2.27) shows that the dependence of mass on radius depends upon whether $M_P \gg M_c$ or $M_P \ll M_c$.

For $M_P \gg M_c$, $M_P R_P^3$ = constant, a relationship met with in the study of white dwarf stars. On the other hand, if $M_P \ll M_c$, $M_P R_P^{-3}$ = constant, and this is the relationship typical of planetary bodies. We can then draw the conclusion that a planet is a particular form of a cold body where the radius increases with the mass. In this way it follows that the magnitude of the critical mass M_c calculated from equation (2.29) is the maximum that can be associated with a planetary body. The arguments confirm those developed in §2.4.

2.7.3 COMMENT ON THE CASE $M \sim M_c$

The arguments developed so far imply a continuous link between cold bodies of small mass and large mass through the condition of maximum radius but this is not observed to be the case. The reason is that our analysis has not accounted for a very important physical process which becomes operative when M_P is only slightly greater than M_c. This is the occurrence of thermonuclear processes when a heat source appears inside the body, which radically affects the equilibrium. Thermal energy now replaces the elastic energy as the counter to gravitational attraction; the body has become a luminous star and so has ceased to be a cold body. Only when the nuclear burning has ceased is it possible for the body to be regarded again as cold, but achieving the 'cold' equilibrium from the 'hot' can involve some catastrophic rearrangement of mass (for instance, the ejection of matter) unless the mass is only marginally sufficient to allow the nuclear burning to occur.

This means in practice that there is not a continuous mass range for a cold body and conditions above M_c can be achieved only after additional physical processes have ceased to act.

2.8 TEMPERATURE RISE ON COMPRESSION

The compression of material causes its temperature to rise. Although a planetary body is cold in the sense already made clear, the temperature will increase inwards due to compression and other causes. We shall consider these matters later, in chapter 7, but it is useful to anticipate this now.

For the compression of a gas we can apply the virial theorem, comparing the thermal energy E_T with the gravitational energy E_G. Taking

$$E_T = 3k\,\Delta T$$

where ΔT is the temperature rise due to compression, we write

$$3k\,\Delta T \sim \frac{1}{2}\frac{GM_P^2}{R_P}\frac{1}{N_p}$$

$$= \frac{1}{2}\frac{GM_P}{R_P}Am_p. \qquad (2.31)$$

Rearranging this expression and introducing the numerical magnitudes for m_p, G and k gives

$$\Delta T \sim 1.35 \times 10^{-15} \frac{M_P}{R_P} A$$

as the temperature rises. For a sphere of hydrogen ($A = 1$) and of maximum size we find $\Delta T = 2.75 \times 10^4$ K. The data for the Jovian planets (presumed to have condensed from gas without heat loss) are collected in table 2.3.

Table 2.3 *Estimates of the internal temperatures of the major planets on the basis of simple compression without heat loss.*

Body	A	T
Jupiter	1	3.56×10^4
Saturn	2	2.56×10^4
Uranus	5	2.25×10^4
Neptune	5	2.73×10^4

This approach may not be adequate for the terrestrial planets whose formation was probably associated with an initial condensation of matter into small flakes no larger than a centimetre or so in radius before consolidation into a planetary object.

2.9 CONCLUSIONS

1. The internal conditions of a body are not dependent on whether it orbits the Sun or a planet, so it is convenient to introduce the concept of a planetary body to include both planets and their satellites.
2. In a planetary body the inward force of gravity is balanced by the resistance of the material to compression. The equilibrium condition is reached when the force of gravity is balanced by the force of material incompression.
3. The strength of incompression is determined by the atomic properties of the constituent matter.
4. There is a largest and smallest mass that a planetary body can have. The largest mass is determined by the condition that the central pressure is sufficiently great to cause the atomic structure to break down. The smallest mass is that where the compression forces are just called into play to

oppose gravity. For smaller bodies the gravitational force is too weak to affect the action of chemical forces.

5 A planetary body will assume a shape of symmetry, given enough time. Smaller bodies will not.

6 The largest possible planetary body is only slightly more massive than Jupiter; the smallest is about the size of the largest asteroids. This corresponds to sizes in the range 10^8 to 10^5 m.

7 The equilibrium characterising a planetary body does not depend on the thermal energy. This can be expressed by saying that the planetary body is cold. Other cold bodies include white dwarf and neutron stars; a planetary body is a member of a wider class of object where gravity is balanced by atomic or nuclear forces.

8 Because the thermal energy is negligible there is an upper limit to the internal temperature to be associated with a planetary body. This must always be less than 10^5 K, and often appreciably less.

9 The internal conditions are such that the material behaves as a very viscous liquid over the majority of the volume. The viscous liquid behaviour will not apply to the surface region because there the gravitational force (manifested through the material pressure) is too low. The internal structure must, therefore, involve at least the two separate regions of crust and interior.

REFERENCES AND COMMENTS

RC2.1 PLANETARY BODY

The general tendency of gaseous material to condense was recognised by Sir James Jeans who showed what the general condition for condensation is in a simple argument. The condition is that the gravitational binding energy for the mass of material shall be larger than the kinetic energy which opposes it. For a mass M of material of radius R, the potential energy E_g of gravitation is, to an order of magnitude,

$$E_g = -GM^2/R.$$

The internal (kinetic) energy E_k for the volume V is

$$E_k = pV$$

where p is the pressure inside the material volume. Gravitational binding is favoured if

$$E_g \gg E_k$$

i.e. if

$$GM^2/R \gg pV.$$

The density ρ of the material is given in terms of M by

$$M = \tfrac{4}{3}\pi\rho R^3$$

so that condensation will occur if

$$GM^2 \gg p(M/\rho)^{4/3}.$$

This can be written in the form $M \gg M_J$ where M_J is the Jeans mass given by

$$M_J = (3/4\pi)^{1/2}(p^{3/2}/\rho^2 G^{3/2}).$$

For a gas obeying the ideal equation of state at temperature T

$$pV = NkT$$

where there are N particles and k is the Boltzmann constant. This allows us to conclude that

$$M_J \approx \rho^{-1/2}T^{3/2}.$$

The Jeans mass decreases with the temperature and can become small if the temperature is low enough. Any planetary mass can be regarded as in excess of an associated Jeans mass provided the pressure and density of the initial gaseous material are suitably arranged and the initial temperature is low enough.

Although the arguments developed so far apply to a gas they can be applied to a collection of small grains of material, regarded as a gas; again condensation will occur if the temperature is low enough. The large mass of each grain restricts its kinetic motion to the extent that it will behave as a gas particle at very low temperature.

RC2.2 COMPRESSION AND THE BULK MODULUS

That the density and compression characteristics of all materials of geophysics are broadly the same follows from very simple arguments.

Consider an atom of mass M and radius $r_a = fa_0$ where a_0 is the Bohr radius and f is a numerical factor close to unity. Suppose the atom to contain A nucleons so that $M = Am_p$. The atomic density ρ_a is

$$\rho_a = 3Am_p/4\pi(fa_0)^3.$$

The atomic radius increases with A and we can represent this dependence to sufficient accuracy for our present purposes by the form $f = f_0A^n$, where f_0 and n are numbers to be assigned. This means

$$\rho_a = (3m_p/4\pi f_0^3 a_0^3)A^{1-3n}.$$

Although the dependence of atomic size on A is not as simple and well behaved as suggested here it is possible to assign $n=0.2$ to get a generally reliable average dependence. For hydrogen ($A=1$), $f_0=1.4$, so we have

$$\rho_a=1.2\times10^3A^{0.4}\ \mathrm{kg\,m^{-3}}.$$

For the naturally occurring elements $1<A<250$ which gives the density range $10^3<\rho_a<10^4\ \mathrm{kg\,m^{-3}}$. These are upper limits because no account has been taken of the packing of atoms in space. We see, however, that the densities of all the elements lie within a factor 10 of $10^3\ \mathrm{kg\,m^{-3}}$.

The compressibility of the material is determined ultimately by the internal kinetic pressure of the atom and this again is fairly constant from one atom to another. Consequently we would expect the compressibility (or its inverse, the bulk modulus K) to be very largely similar in magnitude for all solid material.

To find this magnitude we can consider a metal as a representative solid material and realise that the structure of ions and free (valence) electrons can be treated in the same way as the atom was treated in §2.3. The bulk modulus is defined by

$$K=\rho\ \partial p/\partial\rho$$

and equation (2.13a) will apply for the kinetic pressure. The result is the rough relationship $K\sim p_k\sim10^{11}\ \mathrm{Nm^{-2}}$, having the same dimensions as the pressure. This will be the general order of magnitude of the bulk modulus for solid planetary materials. This is a very rough estimate but does suggest the dependence of planetary density with depth is likely to depend somewhat critically on the value of K, possibly because K will appear in the formulae to some high power.

RC2.3 ATOMIC COMPOSITION

The development of the theory of the atom in the main text is rather cavalier although the results are the same as those derived from the exact theory. Reference to the exact theory can be made using a wide variety of books; a good recent text to be recommended is

Bransden B H and Joachain C J 1983 *Physics of Atoms and Molecules* (London: Longman)

The arguments of the main text can also be applied in general terms to nuclei containing several nucleons. There will be two differences, however, in that the nucleon mass is about 1836 times that of the electron and the nuclear forces are stronger than the electric forces by a factor of about $10^{1/2}$ or 3.16. The formulae (2.7) and (2.10) of the main text still apply, giving a nuclear size of order 10^{-14} m and a characteristic nuclear energy about 10^6 eV. The nuclear internal kinetic pressure follows from equation (2.13a) to be some 10^{19} atmospheres, which is far and away higher than anything met with in planetary materials. Nuclei are entirely stable within planetary bodies; the same is true within most stellar systems (for instance, the central pressure in the Sun is about 10^9 atm). It should be noticed that we are speaking here of the resistance to crushing: the rearrangement of nucleons due to collisions (thermonuclear processes) is a different matter.

We might notice in passing that if the conditions of pressure were ever reached where the nuclei were crushed there would be no forces to resist the gravitational collapse, other than the nuclear constituents (presumably quarks). This is probably the condition in a black hole.

RC2.4 MAXIMUM AND MINIMUM SIZE

Just as there is a maximum size for a planetary body so there is a maximum size for a star. The conditions inside the main body of a star are those of complete ionisation, and the material can be supposed to be interwoven electron and ion gases. The ion gas provides the stellar mass while the electron gas provides the kinetic pressure resisting gravitational compression.

A star of mass M_S will contain $N_p = M_S/m_p$ protons and there will be the same number of electrons. If R_S is the radius of the star the number of particles in it per unit volume is

$$n = 3 \times 2N_p/4\pi R_S^3.$$

The equation of state for the material is that of an ideal gas, so that the gas pressure p_g is given by

$$p_g = 6N_p kT/4\pi R_S^3,$$

where k is the Boltzmann constant and T the temperature.

There is also an outward pressure p_r due to radiation expressed by the Boltzmann–Stefan formula

$$p_r = aT^4/3$$

where a is the radiation constant. If the ratio $P = p_g/p_r > 1$ the gas pressure dominates the radiation pressure; if $P < 1$ the radiation pressure dominates. For the Sun, $M_S = 2 \times 10^{30}$ kg, $R_S = 6 \times 10^6$ m and $T = 5 \times 10^6$ K, from which it follows that $P = 10^4$ and the gas pressure dominates. The same is true for white dwarf and red giant stars. As the mass of the star increases so does its inner temperature and the radiation pressure increases as the fourth power of the temperature. The radiation photons are not effectively attracted by the gravitational force so the kinetic (radiation) energy increases more quickly than does the potential (gravitational) energy. There will come a mass where the two energy contributions are comparable and the total energy (positive kinetic and negative potential) is zero. Further increase of mass will lead to a positive total energy which means the material is not bound. The star cannot hold together and this provides an upper limit for the mass that can be associated with a star. Detailed calculations show this upper mass to be about $60M_{\odot}$ ($\odot$ indicates the Sun).

There is also a clear lower mass, and in this respect a star differs from a planetary body. For a collection of atoms to form a star the internal pressure must be sufficiently large to disrupt the atoms. If the star is to ignite, the inner temperature must also be high enough for thermonuclear processes to occur. This lower energy is known from quantum mechanics to be about 1.23×10^5 eV and this implies a mass for the total body of about $0.1M_{\odot}$. This will be the lowest mass. These various matters can be pursued further in

McCrea W H 1950 *Physics of the Sun and Stars* (London: Hutchinson)
Tayler R J 1970 *The Stars: their structure and evolution* (London: Wykeham)
Kaplan S A *The Physics of the Stars* (New York: Wiley)
Weisskopf V F 1975 Of Atoms, Mountains and Stars: A Study in Qualitative Physics *Science* **187** 605–12

For a short but useful comment on white dwarf stars see

Huang K 1963 *Statistical Mechanics* (New York: Wiley)

The arguments can be extended in various interesting directions but this is outside the scope of our present discussion.

RC2.5 MASS AND RADIUS

Initial discussion of the relation between mass and radius for a range of bodies was made by

Kothari D S 1936 *Mon. Not. R. Astron. Soc.* **96** 833–43

The mass–radius data for several white dwarf stars are shown in figure 2.2. The decrease of radius with mass is readily seen and the dependence follows the prediction of the main text. The essential dynamical difference between a large planetary body of hydrogen and helium and a white dwarf star, primarily composed of helium, is the velocity regimes of the electrons. While the speeds are non-relativistic for a planetary body (where the temperature is less than 10^5 K) they are fully relativistic for a white dwarf (where the temperature is some 10^7 K). This difference will affect the equation of state of the electrons and in particular will provide an upper limit to the strength of opposition to gravity. A distinct upper magnitude of the possible mass for a white dwarf (the Chandrasekhar limit, $= 1.4M_{\odot}$) is an immediate consequence.

The same arguments can be applied to neutron stars. Once the thermonuclear fuel has been used up and the burning has ceased there is no radiation pressure to overcome the gravitational force other than the inherent strength of the nucleons themselves. The electrons and protons combine to form neutrons and the mass collapses until the internal neutron pressure, acting on

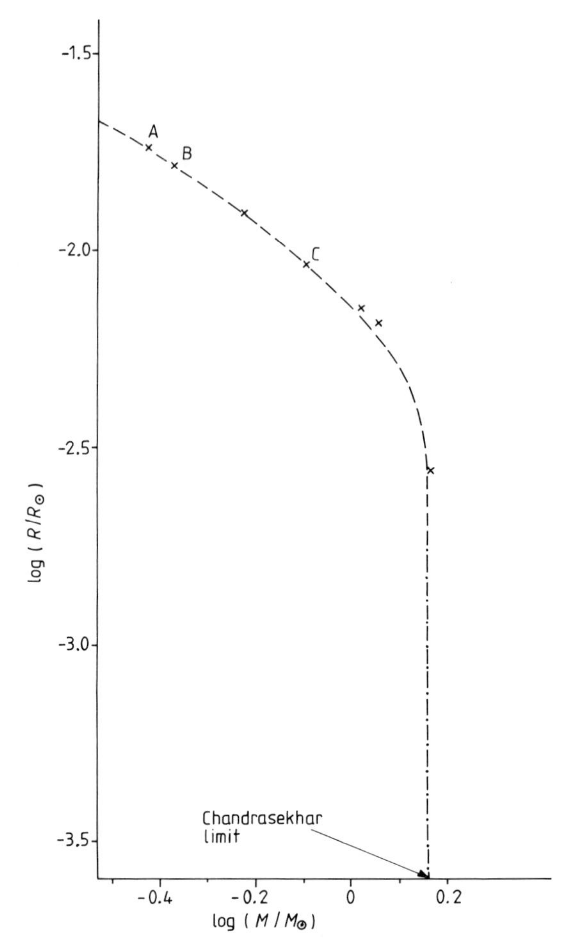

Figure 2.2 *Schematic plot of the radius* R *against mass* M *for seven white dwarf stars relative to the radius* $R_\odot$ *and the mass* $M_\odot$ *of the Sun, in logarithmic form. The upper (Chandrasekhar) limit for the mass,* $\log M/M_\odot = 0.158$ *is clearly evident. Point* A *represents van Moanen 2,* B *represents 20 Eridana B and* C *represents Sirius B.*

contiguous neutrons, can effectively oppose the gravitational contraction. Now gravity is balanced by nuclear forces and the star is essentially a large nucleus with nuclear density. For a solar mass this will mean a radius of the order of 10 km. The outer regions of the neutron star will be different, in that a crustal region can be identified, as for a planet, but the thickness now will be very small. Once again the radius will be predicted to decrease with the increase of mass, roughly according to the inverse $\frac{1}{3}$ power law as for white dwarfs, but possibly modified to some extent because of the particular nature of the short-ranged nuclear forces.

RC2.6 ESTIMATION OF BOUNDS ON PHYSICAL QUANTITIES

The general order-of-magnitude approach to estimating the magnitudes of physical quantities can be taken a stage further by deducing upper and lower limits (bounds). For a planetary interior this can be based on the assumptions of hydrostatic equilibrium (see chapter 3) with an associated monotonic increase of density throughout the planetary body. Of interest are central pressure and density and surface density. These matters have been considered and summarised by

de Marcus W C 1958 *Encyclopaedia of Physics* (Berlin: Springer) pp 419–48

RC2.6.1 Central pressure and density. The actual central pressure P_c for a body of given mass and radius must be larger than it would be if the density were uniform throughout. The actual central density will be less than it would be if the structure of the body were a uniform density of magnitude equal to that of the central pressure but extending only over a radius $R_1 < R_P$. These conditions provide a lower bound for the pressure in the notation of chapters 3 and 4 of the form

$$P_c > \tfrac{1}{2}(4\pi/3)^{1/3} G M_P^{2/3} \hat{\rho}^{4/3}(1-2w),$$

where $\hat{\rho}$ is the mean density of the body and w is the ratio of the centripetal to gravitational forces at the equator.

For Jupiter this implies a central pressure in excess of 1.1×10^{12} N m^{-2} while for Saturn the magnitude is 2.1×10^{11} N m^{-2}. The atomic kinetic pressure, it will be remembered, is about 3×10^{12} N m^{-2}, showing again the nearness of Jupiter to the maximum planetary conditions.

RC2.6.2 Weighted average pressure. Consider the pressure $\langle P \rangle$ weighted according to mass and defined by

$$\langle P \rangle = \frac{1}{M_P} \int^{M_P} P \, \mathrm{d}M$$

$$= -\frac{1}{M_P} \int_{M_P} M \frac{\mathrm{d}P}{\mathrm{d}M} \, \mathrm{d}M.$$

For hydrostatic equilibrium the arguments of chapter 3 lead to the condition

$$\mathrm{d}P/\mathrm{d}M = -GM/4\pi R^4 + w^2/6\pi R$$

for the distance R from the centre. This leads to the expression for $\langle P \rangle$ which can be arranged to account for the condition that the density never decreases with distance from the centre. The result is the condition

$$\langle P\rangle > \tfrac{1}{5}(4\pi/3)^{1/3} G M_{\mathrm{P}}^{2/3} \hat{\rho}^{2/3}(1-2w/3).$$

According to this expression the *average* pressure within Jupiter must exceed 4.6×10^{11} N m^{-2} if the monotonic density decrease from the centre is to be maintained.

RC2.6.3 Surface density.

Jeffreys 1954 *Mon. Not. R. Astron. Soc.* **14** 433

In this, it was shown that the surface density ρ_s for a planetary body must be less than a magnitude dictated by the inertia factor α_{P} representing the distribution of material in the body. Explicitly:

$$\rho_s < 5\hat{\rho}\alpha_{\mathrm{P}}/2. \qquad \text{(RC2.1)}$$

This condition has been refined by de Marcus (see reference above) by introducing the coefficients J_n in the expansion of the gravitational field (see chapter 4). Explicitly, he shows that including J_2

$$\rho_s < (5J_2/3)(J_2 + w/2)\hat{\rho}, \qquad \text{(RC2.2)}$$

while, including both J_2 and J_4,

$$\rho_s < [7J_4/(2J_2 + w)(J_2 + 3w) + 7J_4/3]\hat{\rho}. \qquad \text{(RC2.3)}$$

These formulae lead to the following maximum surface densities, taken from the de Marcus paper quoted above, collected in table RC2.1. The formulae are applicable within the planetary body as well but the conditions of ellipticity are not necessarily known in that context.

Table RC2.1 *Values of the maximum surface densities for several planets, calculated using the expressions (RC2.1) (Jeffreys 1954), (RC2.2) (de Marcus 1958; J_2) and (RC2.3) (de Marcus 1958; J_2, J_4).*

Body	Maximum surface density (kg m^{-3})		
	Jeffreys (1954)	de Marcus (1958)	
		J_2	J_2, J_4
Earth	4600	4480	4100
Mars	4010	—	—
Jupiter	860	760	570
Saturn	370	330	—
Neptune	1850	1780	—

RC2.7 THE VIRIAL THEOREM

This is a very general and powerful statement relating the mean values of the kinetic and potential energies of a collection of interacting particles contained in a finite region of space. The mean value of a quantity A, written $\bar{A}$, is the time average over an interval t to $t+\tau$ and is defined by

$$\bar{A}=\lim_{\tau\to\infty}(1/\tau)\int_t^{t+\tau}A(t)\,\mathrm{d}t.$$

Consider a collection of N particles, the jth particle having mass m_j and located at the position x_j with speed $\dot{x}_j$. The kinetic energy K of the particles is

$$K=\sum_{j=1}^{N}\tfrac{1}{2}m_j\dot{x}_j^2=\sum_{j=1}^{N}\tfrac{1}{2}p_j\dot{x}_j$$

$$=\tfrac{1}{2}\sum_{j=1}^{N}[\mathrm{d}/\mathrm{d}t(p_jx_j)-x_j\dot{p}_j].$$

From Newton's second law of motion

$$\dot{p}_j=-\mathrm{d}V/\mathrm{d}x_j$$

where $V(x_j)$ is the potential energy of the system. Then, because p_j and x_j remain finite always,

$$\overline{\mathrm{d}/\mathrm{d}t(p_jx_j)}=0.$$

Forming the mean value of the kinetic energy we have

$$2\bar{K}=\overline{\left(\sum_j x_j\,\mathrm{d}V/\mathrm{d}x_j\right)}.$$

The potential energy is a homogeneous function of degree k so we have $V\propto x^k$ and

$$\sum_j x_j\,\mathrm{d}V/\mathrm{d}x_j=kV.$$

For gravitation or for electrostatic interactions $k=-1$ so that

$$2\bar{K}=-\bar{V}.$$

This is an expression of the virial theorem and is central to many of our arguments.

3

CONSEQUENCES OF INTERNAL PRESSURE: HYDROSTATIC EQUILIBRIUM

A planetary body is, according to the specification in the last chapter, a body where the transverse components of the interatomic forces play no role in determining the overall shape of the body. Thermal energy plays no part in establishing or maintaining the equilibrium. This means that the material has negligibly small rigidity on a sufficiently long time scale and so shows the properties of slow flow. Conditions are isothermal as far as the flow characteristics are concerned. Several consequences follow from this particular situation and we investigate these matters in the present chapter.

3.1 HYDROSTATIC CONDITIONS

The condition of material equilibrium in a planetary body is that of hydrostatic equilibrium. The expression of this equilibrium in a particular case depends on the laws of fluid statics supplemented by Archimedes' law of buoyancy.

The laws of fluid statics can be expressed for our present purposes by two statements. First, the pressure is the same at every point in the continuous layer of fluid at the same distance from the centre of mass of the material volume. Second, the pressure in a given layer of a homogeneous fluid is greater than the pressure at a layer further from the centre by an amount numerically equal to the weight of the material between the two layers. The immediate consequence of these laws is that a self-gravitating fluid not in rotation will assume the figure of a sphere. This must be the case, for otherwise the pressure at the centre would be different in different directions because the amount of material between the centre and the surface would depend on the direction (that is latitude and longitude). The pressure is a scalar quantity and so has a magnitude independent of direction. This can be so only if the shape is spherical, latitude and longitude being irrelevant.

For a rotating fluid the shape is distorted by the force of rotation. This has its maximum effect at the equator and is zero at the poles, marked where the hypothetical rotation axis emerges from the body. The rotation force acts outwards, away from the centre and so opposes the gravitational force which draws the body together. The compression of the material in the equatorial direction is therefore less than it would be in the absence of rotation, the equatorial radius R_E being greater than the simple spherical radius. The effect of rotation diminishes as the cosine of the latitude (see §3.2) and vanishes at the poles. This does not mean that the polar radius R_P is unaffected by rotation. The amount of material in the planetary body is constant so the volume associated with the latitude distortion must be found elsewhere and this must involve the polar regions. Equatorial gain must mean polar loss and the polar radius will decrease as a consequence. The spherical shape ($R_E = R_P$) of a non-rotating fluid is changed by rotation into the shape of a spheroid of revolution ($R_E > R_P$) with symmetry about the rotation axis. The figure of a cross section containing the polar and equatorial direction is an ellipse of semi-major axis R_E and semi-minor axis R_P. The relative difference between the equatorial and polar radii defines the ellipticity e, sometimes called the polar flattening, according to

$$e = (R_E - R_P)/R_E. \quad (3.1)$$

The ellipticity depends on the angular velocity w of rotation, and we will be concerned with the details of this dependence in chapter 4. The shapes taken up by rotating fluid masses have been investigated by many authors following the work of Jacobi and Maclaurin. As the angular speed increases so the difference increases and ultimately the stability of the mass is lost. The condition of instability lies outside the range of conditions met within planetary bodies associated with the Solar System, at least at this stage of its evolution.

Conditions within a fluid body are described by Archimedes' law of buoyancy. The fluid exerts a net upward force on a body immersed in it of magnitude equal to the weight of the fluid displaced by the body. This upward force exerted by the fluid is the buoyancy force. The resultant force on the immersed body is the difference between the gravitational force downwards and the buoyancy force acting upwards. If the immersed body has a mass less than that of the displaced fluid the buoyancy force dominates and the body tends to rise; alternatively, if the mass of the body is greater than that of the displaced fluid the body tends to sink.

These results apply to fluids and are independent of the time necessary to reach the equilibrium state. This being the case, material with very high viscosity showing slow flow can also take up the same equilibrium shape. Therefore, the main bulk of a planetary body will have an equilibrium shape of the fluid form because the material below the thin crust shows slow flow.

The buoyancy effects also have an important consequence. The lighter material will rise and the heavier material will consequently fall, with the result that the more dense material will lie at the centre of the planetary body. The distribution of chemical materials will not be homogeneous, the material being said to be differentiated. The heavier material must concentrate towards the centre of the planetary body as a final equilibrium state. Whether this state will be achieved will depend on the rate at which the differentiation processes take place. We shall consider this further in §3.2.

3.2 DIFFERENTIATION OF MATERIALS

An immediate consequence of the hydrostatic conditions is the separation of materials of different density, the light materials rising and the heavier materials falling within the body. For a continuous spectrum of materials this would lead ultimately to a continuous increase of material density with depth, quite separately from that associated with the effects of compression. These effects will be more marked for the discontinuous range of materials actually found in the Solar System. We saw in chapter 1 that the materials of importance in planetary studies are hydrogen and helium, silicates and free ferrous compounds, and here the densities are very clearly separated in magnitudes. For instance, silicates will generally have densities in the range 2500–4000 $kg\,m^{-3}$ at zero pressure, whereas the free ferrous materials have densities in excess of 7000 $kg\,m^{-3}$. For a mixture of elements of cosmic composition (see table 1.2) arranged in this molecular form, the separation will be clear. The free iron, cobalt, nickel and ferrous mixtures will sink to the central region, so forming a core, while the lighter silicates will form an enclosing mantle again structured by density. Lighter elements (predominantly hydrogen and helium) will be above this again. If the proportion of hydrogen and helium is large the internal pressure may be sufficient to produce high density condensed versions of these elements.

For a body of cosmic composition we can refer to the data of chapter 1 to obtain an overall indication of the composition. The hydrogen and helium component will collectively account for some 98% by weight (see table 1.2). Of the remaining 2%, about 21% will be free ferrous material, on the basis of table 1.9 for chondritic material, which is about 0.0042% of the whole. For a body of mass 10^{27} kg, as an example, this would imply a central ferrous component of mass about 4×10^{24} kg (roughly the Earth's mass) surrounded by a silicate component of about 1.96×10^{25} kg to form a central core of some 3.3 Earth masses. The remaining mass, comprising a mantle and an atmosphere, would be predominantly hydrogen and helium and of some 164 Earth masses. Such a body might be expected to have properties closely similar to those of Jupiter (mass 1.9×10^{27} kg) or Saturn (mass 5.7×10^{26} kg) but we shall see in chapter 6 how this expectation is not entirely met by the observation of these planets. The basic idea of differentiation of material is, however, borne out in practice both for the major planets and for the terrestrial planets and it is this feature that particularly concerns us now.

Having concluded that differentiation of the material can be expected to occur under conditions of hydrostatic equilibrium, it is important to gain an estimate of how long would be necessary for such separation to come about starting with a body of a complete mix of materials. It is possible, of course, that the differentiation occurred at the time of formation of the body, in which case the problem would not occur.

Assume the body to consist of a complete mix of materials initially and suppose the principal separation is that between the silicate and free ferrous materials. Consider the idealised case of a small spherical volume V_1 of material density ρ_1 in a medium of density $\rho < \rho_1$. Because the small mass is more dense than the surroundings it will sink under local gravity. As it sinks it is retarded by viscous friction. A steady state is reached, with motion at constant speed, if the moving force F_G due to gravity is exactly balanced by the viscous (friction) force F_v. For this case we have $F_G = F_v$. If the viscous force follows Stokes's law we write, for a sphere of radius a moving with speed $\bar{v}$:

$$F_R = 6\pi\eta a\bar{v} \qquad \text{and} \qquad F_G = (\rho_1 - \rho)V_1 g \tag{3.2}$$

where η is the coefficient of shear viscosity of the medium. Then

$$\rho(\rho_1/\rho - 1)\tfrac{4}{3}\pi a^3 g = 6\pi\eta a\bar{v}. \tag{3.3}$$

It is a result of fluid mechanics that Stokes's law is applicable provided the Reynolds number is small, say less than about $\frac{1}{10}$, so that

$$\rho\bar{v}a/\eta < \tfrac{1}{10}. \tag{3.4}$$

We shall find that this condition is well satisfied under planetary conditions, primarily because the viscosity η is very high.

Writing $y = \rho_1/\rho$, and with $\nu = \eta/\rho$ being the kinematic viscosity, equation (3.3) provides the expression for the velocity

$$\bar{v} = \frac{2a^2 g}{9\nu(y-1)}. \tag{3.5}$$

The acceleration due to gravity g is written in terms of the mass M contained within the spherical volume of radius R as

$$g = \frac{GM}{R^2} = \frac{G}{R^2} \int_0^R \rho(r) r^2 \, dr.$$

The speed $\bar{v}$ can be interpreted as the time t_L for the sphere to fall through a distance L. From equation (3.5) we find

$$t_L \sim \frac{9v(y-1)L}{2a^2 G} \left(\frac{1}{R^2} \int_0^R \rho(r) r^2 \, dr \right)^{-1}. \tag{3.6}$$

For free iron falling through silicates we have $y \sim 2$ so that $(y-1) \sim 1$. The time t_L is seen to decrease as a increases but increases as v increases. Indeed, the crucial factor for a given body is the magnitude of the kinematic viscosity.

For a sphere of radius 1 m, equation (3.5) reduces to

$$\bar{v} \sim (g/v)10^{-1}.$$

For terrestrial planetary material, $v \sim 10^{17}$ m^2 s^{-1} so that $\bar{v} \sim 10^{-18} g$. For the mantle of the Earth $g \sim 10$ so that $\bar{v} \sim 10^{-17}$ m s^{-1}. The condition (3.4) is amply satisfied. The time t_L for the sphere to pass through the mantle ($L \sim 3 \times 10^6$ m) is then $t_L = 10^{27}$ s. This time will fall as the magnitude of kinematic viscosity decreases, for instance as the result of a rise of local temperature. The reduction of the viscosity by a factor 10^{12} due to temperature increase will reduce the time for differentiation to 10^{15} s which is of the order of 10^8 years. The effect of temperature on the viscosity is seen to be very important. Although the thermal energy plays no role in determining the equilibrium overall it is important in determining the details of the internal structure.

3.3 INCREASE OF DENSITY WITH DEPTH

The change of density with depth is related to the compression of the material.

3.3.1 NO ROTATION

The simplest case is that of a spherically symmetric chemically homogeneous fluid sphere not in rotation. The density will depend only on the distance from the centre, the internal conditions being isotropic. The pressure difference dp between two layers within the body with radii R and $R + dR$ is proportional to the separation dR: explicitly

$$dp = -\rho(R) g(R) \, dR \tag{3.7}$$

where ρ and g refer to the radius R. If $M(R)$ is the mass of material contained within the sphere of radius R and G is the universal constant of gravitation:

$$\begin{aligned} g(R) &= GM(R)/R^2 \\ &= (4\pi G/R^2) \int_0^R \rho(r) r^2 \, dr. \end{aligned} \tag{3.8}$$

Combining equations (3.7) and (3.8):

$$\frac{\mathrm{d}p}{\mathrm{d}R} = -(4\pi G/R^2)\rho \int_0^R \rho(r) r^2 \, \mathrm{d}r. \tag{3.9}$$

Because thermal energy is neglected, conditions are taken to be isothermal. If the density is a constant ρ_0 throughout the body, equation (3.9) reduces to

$$\frac{\mathrm{d}p}{\mathrm{d}R} = -\tfrac{4}{3}\pi G \rho_0^2 R.$$

The pressure p_c at the centre is

$$p_c = \tfrac{2}{3}\pi G \rho_0^2 R^2.$$

The expansion (3.9) for the pressure gradient is readily converted into a corresponding expression for the gradient of the density, for

$$\frac{\mathrm{d}p}{\mathrm{d}R} = \frac{\mathrm{d}p}{\mathrm{d}\rho}\frac{\mathrm{d}\rho}{\mathrm{d}R}.$$

Using equation (2.2) this is written

$$\frac{\mathrm{d}p}{\mathrm{d}R} = \frac{K}{\rho}\frac{\mathrm{d}\rho}{\mathrm{d}R}$$

where K is the bulk modulus of the material. Equation (3.9) is consequently converted into the expression

$$\frac{\mathrm{d}\rho}{\mathrm{d}R} = -\frac{4\pi G}{K}\frac{\rho^2}{R^2}\int_0^R \rho(r) r^2 \, \mathrm{d}r. \tag{3.10}$$

The condition of constant density throughout would be possible only if compressibility were zero and so the bulk modulus indefinitely large. For a real material, K will generally be of the order of 10^{11} N m^{-2} for pressures up to about the same magnitude.

If the bulk modulus is known as a function of the depth, which means also of the pressure, equation (3.10) provides a way of finding the density profile through the body. If R_p is the total radius and M_p the total mass, the gradient of the density at the surface is given by

$$\left.\frac{\mathrm{d}\rho}{\mathrm{d}R}\right|_{R_P} = -\frac{4\pi G}{K}\frac{\rho^2}{R_P^2}\int_0^{R_P} \rho(r) r^2 \, \mathrm{d}r \tag{3.11}$$

and

$$M_P = 4\pi \int_0^{R_P} \rho(r) r^2 \, \mathrm{d}r. \tag{3.11a}$$

The density at the depth dR below the surface is written

$$\rho(R_P - dR) = \rho(R_P) + \left.\frac{d\rho}{dR}\right|_{R_P} dR.$$

The insertion of this density into equations (3.11) and (3.11a) gives the gradient

$$\left.\frac{d\rho}{dR}\right|_{R_P - dR} = -\frac{4\pi G}{K}\frac{\rho^2(R_P - dR)}{(R_P - dR)^2}\int_0^{R_P - dR} \rho(r) r^2 \, dr$$

$$M(R_P - dR) = 4\pi \int_0^{R_P - dR} \rho(r) r^2 \, dr.$$

The density at the depth $R_P - 2dR$ is written

$$\rho(R_P - 2dR) = \rho(R_P - dR) + \left.\frac{d\rho}{dR}\right|_{R_P - dR} dR.$$

The computation can be continued to the centre to provide the full density profile with depth. The data are also sufficient to give the pressure profile using equation (3.9). The computation is tedious for a body of planetary size but is practicable using electronic computation techniques.

The calculation cannot be applied directly to actual planetary bodies because the bulk modulus of material K is not known directly. We shall see in chapter 5 that this information has been inferred for the Earth using seismic measurements and when used with equation (3.10) provides a method of determining ρ/K as a function of the depth. The approach was first suggested by Adams and Williamson and is now a well established way of exploring the interior of the Earth. This approach is available for the other terrestrial planets once the necessary seismic data are available. We will have more to say about this in chapter 8.

3.3.2 INTERNAL INHOMOGENEITIES

The method presumes chemical homogeneity but the arguments of §3.2 lead us to expect the interior of a planetary body to be differentiated. The density will be a continuous function of the depth only in restricted regions because a region of homogeneity will form a spherical shell about the centre. The arguments of Adams and Williamson will apply within the shell and the procedure will be applied separately from one shell to the next, moving inwards from the surface.

Let us introduce the function ϕ by

$$\phi = K/\rho$$

so that

$$dK = \rho \, d\phi + \phi \, d\rho.$$

This means that

$$\frac{\mathrm{d}K}{\mathrm{d}p}=\rho\frac{\mathrm{d}\phi}{\mathrm{d}p}+\phi\frac{\mathrm{d}\rho}{\mathrm{d}p}$$

$$=\rho\frac{\mathrm{d}\phi}{\mathrm{d}z}\left(\frac{\mathrm{d}p}{\mathrm{d}z}\right)^{-1}+\phi\frac{\mathrm{d}\rho}{\mathrm{d}z}\left(\frac{\mathrm{d}p}{\mathrm{d}z}\right)^{-1}$$

where z is the depth below the surface. The hydrostatic condition is

$$\mathrm{d}p=\rho g\,\mathrm{d}z$$

so that

$$\frac{\mathrm{d}K}{\mathrm{d}p}=\frac{\phi}{\rho g}\frac{\mathrm{d}\rho}{\mathrm{d}z}+\frac{1}{g}\frac{\mathrm{d}\phi}{\mathrm{d}z}.$$

This means that for the density

$$\frac{\mathrm{d}\rho}{\mathrm{d}z}=\frac{\rho g}{\phi}\left(\frac{\mathrm{d}K}{\mathrm{d}p}-\frac{1}{g}\frac{\mathrm{d}\phi}{\mathrm{d}z}\right).$$

This can be written

$$\frac{\mathrm{d}\rho}{\mathrm{d}z}=\zeta\frac{\rho g}{\phi} \tag{3.13a}$$

where

$$\zeta=\frac{\mathrm{d}K}{\mathrm{d}p}-\frac{1}{g}\frac{\mathrm{d}\phi}{\mathrm{d}z}. \tag{3.13b}$$

For a homogeneous region, $\zeta=1$. The degree to which homogeneity applies can be estimated by the degree to which $(\zeta-1)$ differs from zero. We shall find in chapter 8 that ζ is a sensitive index to chemical inhomogeneities within the Earth.

3.3.3 *ROTATION*

When rotation is taken into account, equation (3.8) must be modified to account for the effects of the centripetal force. For rotation with angular velocity of magnitude w the equation for hydrostatic equilibrium can be written in the same form as equation (3.9) but now involving w. We will see in chapter 4 that, if terms of order w^4 or smaller are neglected, the expression for hydrostatic equilibrium becomes

$$\frac{\mathrm{d}p}{\mathrm{d}R}=-\rho(R)\left(\frac{GM(R)}{R^2}-\tfrac{2}{3}w^2R\right). \tag{3.14}$$

The variable R is now to refer to a spherical surface even though the surfaces of constant density are ellipsoids of revolution. For each surface of constant density within the body, R is assigned the magnitude of the radius of the sphere containing the same volume as the elliptical surface and of mass $M(R)$. The approximation (3.14) is always adequate for the terrestrial planets but is not always so for the major planets. Rotation effects can often be neglected in the treatment of planetary atmospheres. Usually we will set $w=0$.

3.3.4 *A MATHEMATICAL REARRANGEMENT*

The calculation of the change of density with depth using equation (3.10) is an unwieldy task because the unknown density also appears under the integral sign. The equation is called an integro–differential equation and it is always desirable to replace such an equation by an equivalent differential equation which does not involve the integral at all. We do this by differentiating the initial equation with respect to R.

From equations (3.14) and (2.2) we have the expression including rotation with angular speed w:

$$\frac{\mathrm{d}\rho}{\mathrm{d}R}=-\frac{4\pi\rho^2 G}{KR^2}\int_0^R \rho(r)r^2\,\mathrm{d}r+\frac{2}{3}\frac{\rho^2}{K}w^2R$$

so that

$$\int_0^R \rho(r)r^2\,\mathrm{d}r=-\frac{KR^2}{4\pi\rho^2 G}\frac{\mathrm{d}\rho}{\mathrm{d}R}+\frac{w^2R^3}{6\pi G}. \qquad (3.15)$$

The integral is eliminated by differentiating once with respect to R to obtain

$$\rho(R)R^2=-\frac{1}{4\pi G}\left[\frac{KR^2}{\rho^2}\frac{\mathrm{d}^2\rho}{\mathrm{d}R^2}+\frac{2KR}{\rho^3}\frac{\mathrm{d}\rho}{\mathrm{d}R}-\frac{2KR^2}{\rho^3}\left(\frac{\mathrm{d}\rho}{\mathrm{d}R}\right)^2+\frac{R^2}{\rho^2}\left(\frac{\mathrm{d}\rho}{\mathrm{d}R}\right)\left(\frac{\mathrm{d}K}{\mathrm{d}R}\right)\right]+\frac{w^2R^2}{2\pi G}.$$

Rearranging and collecting the terms gives the second-order differential equation

$$\frac{\mathrm{d}^2\rho}{\mathrm{d}R^2}+\left(\frac{2}{R}+\frac{1}{K}\frac{\mathrm{d}K}{\mathrm{d}R}\right)\frac{\mathrm{d}\rho}{\mathrm{d}R}-\frac{2K}{\rho}\left(\frac{\mathrm{d}\rho}{\mathrm{d}R}\right)^2+\frac{4\pi G\rho^3}{K}=\frac{2w^2\rho^2}{K}. \qquad (3.16)$$

Being a second-order equation, two conditions are necessary to obtain a solution of physical interest. One could be the density at the surface, ρ_0. This is not necessarily a useful starting point. From equation (3.10), we see that $\mathrm{d}\rho/\mathrm{d}R$ vanishes at $R=0$. Again, at the centre $\rho=\rho_c$ (for $R=0$). We can select these two conditions as the means of producing the dependence of density on distance. The result will be a magnitude ρ_0 at the distance R_0, the radius of the sphere. With ρ_0 specified, the calculation is repeated for different magnitudes of ρ_c until the correct magnitude of $\rho_0(R_0)$ is obtained. This method gives a value of

R_p as well as of ρ_c. Alternatively, if $\rho_0(R_0)$ and the total mass M_0 are assigned, $(\mathrm{d}\rho/\mathrm{d}R)_0$ at the surface follows from equation (3.11) and the integration of equation (3.16) can be made starting from the surface.

Calculations of this kind assume that K is known with the depth. The variable R appearing in equation (3.16) (see §3.3.3) is not the actual distance from the centre, but the radius of the sphere at each distance having the same volume as the actual shape. The dependence of the pressure on depth follows by integrating equation (3.9). Alternatively, a differential equation analogous to equation (3.16) can be formed for the pressure and solved subject to the boundary conditions that ρ and M are prescribed at the surface.

3.4 EMPIRICAL EQUATIONS OF STATE

The expressions for hydrostatic equilibrium can be developed further if the dependence of bulk modulus on the pressure (and so on the depth) is known. This will involve the dependence of density on the pressure, which is the equation of state of the materials. It is easiest to approach this matter through the bulk modulus.

K will change with the external pressure and we can expect a simple expansion of K in powers of p to represent this change over at least a restricted range of pressure. This means that we write

$$K = K_0 + Bp + Cp^2 + Dp^3 + \cdots \tag{3.17}$$

where K_0 is the bulk modulus at zero pressure and $B, C, D, \ldots$ are coefficients that will depend upon the chemical composition directly, and so on the temperature and pressure indirectly. Remembering equation (2.2) we have

$$\rho \frac{\partial p}{\partial \rho} = K_0 + Bp + Cp^2 + Dp^3 + \cdots \tag{3.18}$$

and this expression can be integrated to give p as a function of ρ, which is the equation of state of the material.

Equation (3.18) can be fitted to experimental data by cutting the series off at a chosen power of the pressure and making a numerical polynomial fit with the given data. It is sufficient for pressures less than about 5×10^{11} N m^{-2} (about 5×10^6 atm) to terminate equation (3.18) at the term involving p^2 and so to set $D = 0$. For pressures up to about 10^{11} N m^{-2} it is usually sufficient, for the purposes of calculation, to set $C = 0$ as well, representing the bulk modulus as a linear function of the pressure for a particular chemical structure. There are three cases that can interest us particularly.

3.4.1 *K INDEPENDENT OF PRESSURE*

In this case, equation (3.18) gives

$$\frac{d\rho}{\rho}=\frac{1}{K_0}\,dp$$

and integration gives

$$\rho=\rho_0\exp\left[(p-p_0)/K_0\right].$$

For small increments of pressure ($p-p_0 \ll K_0$) the exponential can be expanded to provide the dependence of density on the pressure

$$\rho=\rho_0\left(1+\frac{(p-p_0)}{K_0}+\frac{1}{2}\frac{(p-p_0)^2}{K_0^2}+\frac{1}{6}\frac{(p-p_0)^3}{K_0^3}+\cdots\right). \tag{3.19}$$

3.4.2 *LINEAR DEPENDENCE ON PRESSURE*

For higher pressures, equation (3.18) is written

$$\frac{d\rho}{\rho}=\frac{dp}{K_0+Bp}.$$

Integration gives for the pressure

$$p=\frac{K_0}{B}\left[\left(\frac{\rho}{\rho_0}\right)^B-1\right]+p_0\left(\frac{\rho}{\rho_0}\right)^B$$

while the density is expressed by

$$\rho=\rho_0\left(\frac{K_0+Bp}{K_0+Bp_0}\right)^{1/B}.$$

The two formulae simplify if the lower pressure is in fact zero, for then we find

$$p=\frac{K_0}{B}\left[\left(\frac{\rho}{\rho_0}\right)^B-1\right]$$

$$\rho=\rho_0\left(1+\frac{B}{K_0}p\right)^{1/B}. \tag{3.20}$$

This equation of state was first recognised and applied to planetary problems by Murnaghan.

The bulk modulus can be determined directly by differentiation to give

$$K=\left(\frac{\rho}{\rho_0}\right)^B K_0$$

or

$$K/\rho^B=\text{constant}. \tag{3.21}$$

This expression can be treated as an alternative form of the equation of state. For terrestrial materials it is appropriate to choose B in the range $3 \leqslant B \leqslant 5$. The formulae give a satisfactory equation of state for hydrogen and helium with $B \sim 2$.

3.4.3 *QUADRATIC DEPENDENCE ON THE PRESSURE*

For higher pressures, equation (3.18) gives the next approximation

$$\frac{\mathrm{d}p}{K_0 + Bp + Cp^2} = \frac{\mathrm{d}\rho}{\rho}.$$

The form of the integral of this expression depends on the relative magnitudes of K_0, B and C. For terrestrial material, we must take $K_0 > B > C$. Then we integrate to obtain

$$p = \frac{2K_0(\rho^a - \rho_0^a)}{(B+a)\rho_0^a - (B-a)\rho^a} \tag{3.22}$$

and

$$a = (B^2 - 4K_0 C)^{1/2}.$$

This equation of state has been used in studies of the terrestrial planets by Lyttleton. It is considerably more complicated than equation (3.20) for use in the study of planetary interiors.

3.5 MASS AND DENSITY FOR A MURNAGHAN MATERIAL

The differential equation (3.16) requires $K(R)$ to be specified if it is to be applied to a particular case. One choice of interest is that using equation (3.21), because this equation of state is found to represent actual materials well over a wide range of pressures.

3.5.1 *DEPENDENCE OF MASS ON RADIUS*

Consider the mass $\mathrm{d}M$ contained within the thin spherical shell of thickness $\mathrm{d}R$ and constant density ρ. Then

$$\mathrm{d}M = 4\pi R^2 \rho \, \mathrm{d}R. \tag{3.23}$$

The condition for hydrostatic equilibrium then becomes

$$\begin{aligned}\frac{\mathrm{d}\rho}{\mathrm{d}R} &= -\frac{G\rho^2}{KR^2} M + \tfrac{2}{3}\rho^2 \frac{Rw^2}{K} \\ &= -\frac{G\rho_0^B}{K_0} \frac{\rho^{(2-B)}M}{R^2} + \frac{2}{3} \frac{Rw^2 \rho_0^B \rho^{(2-B)}}{K_0}\end{aligned} \tag{3.24}$$

using equation (3.21). Because the effects of rotation are included, the variable R is a hypothetical distance from the centre as explained before. From equation (3.23), the density is given by

$$\rho = \frac{1}{4\pi R^2} \frac{dM}{dR}. \tag{3.25}$$

Differentiating this expression with respect to R gives

$$\frac{d\rho}{dR} = \frac{1}{4\pi R^2} \frac{d^2 M}{dR^2} - \frac{1}{2\pi R^3} \frac{dM}{dR}. \tag{3.26}$$

Equating (3.24) and (3.26), and using equation (3.25), we obtain an equation for M but not ρ directly. On rearrangement, this equation is

$$\frac{d^2 M}{dR^2} - \frac{2}{R} \frac{dM}{dR} + \left(\frac{4\pi^{(B-1)} G \rho_0^B}{K_0} \right) R^{(2B-4)} M \left(\frac{dM}{dR} \right)^{(2-B)} = \frac{2}{3} \frac{(4\pi)^{(B-1)} \rho_0^B}{K_0} R^{(2B-1)} \left(\frac{dM}{dR} \right)^{(2-B)} w^2. \tag{3.27}$$

This rather complicated equation can be conveniently expressed in dimensionless form by introducing the dimensionless variables m and r by

$$M = mM_0 \qquad \text{and} \qquad R = rR_0 \tag{3.28}$$

where M_0 and R_0 are respectively the total mass and radius of the body. For a planetary body this will refer to the volume that remains when the crust is removed from the observed shape.

Inserting equation (3.28) into equation (3.27) gives

$$\frac{d^2 m}{dr^2} - \frac{2}{r} \frac{dm}{dr} + \theta_B r^{(2B-4)} m \left(\frac{dm}{dr} \right)^{(2-B)} = \phi_B r^{(2B-1)} \left(\frac{dm}{dr} \right)^{(2-B)} \tag{3.29}$$

where

$$\theta_B = (4\pi)^{(B-1)} \frac{G \rho_0^B}{K_0} R_0^{(3B-4)} M_0^{(2-B)} \tag{3.29a}$$

and

$$\phi_B = \frac{2}{3} \frac{(4\pi)^{B-1} \rho_0^B}{K_0} R_0^{(3B-1)} M_0^{(1-B)} w^2. \tag{3.29b}$$

The parameters θ_B and ϕ_B describe the magnitude of the mass and effects of rotation respectively. For no rotation we set $\phi_B = 0$. The magnitude of θ_B is determined by the parameters R_0, M_0, K_0, B and ρ_0, while that of ϕ_B involves w as well.

The two conditions necessary to select solutions of physical interest will usually involve the masses at the inner and outer radii of regions of interest.

For a homogeneous sphere this is

$$m(0)=0 \qquad (r=0) \qquad m(1)=1 \qquad (r=1). \tag{3.30}$$

Equation (3.29) is generally a non-linear equation except for the special case $B=2$ when it takes a linear form. This special case is often called the Laplace approximation. The assumption $B=2$ is implicit in an expression for the compression used by Laplace.

3.5.2 *DEPENDENCE OF DENSITY ON RADIUS*

A knowledge of the mass as a function of distance can be converted to that of the density on distance using equation (3.25). The corresponding pressure then follows from equation (3.14). The density can, however, be determined directly by solving a modified form of equation (3.29) well known in the theory of stellar structure, where it is called the Emden equation.

We first transform equation (3.25) into reduced variables using equation (3.28). Then

$$\begin{aligned}\rho &= (M_0/4\pi R_0^3)r^{-2}\,\mathrm{d}m/\mathrm{d}r\\ &= (\hat{\rho}/3)r^{-2}\,\mathrm{d}m/\mathrm{d}r\end{aligned}$$

where

$$\hat{\rho}=3M_0/4\pi R_0^3$$

is the mean density of the complete sphere. This means

$$\frac{\rho}{\hat{\rho}}\equiv\bar{\rho}=\frac{1}{3r^2}\frac{\mathrm{d}m}{\mathrm{d}r} \tag{3.31a}$$

for the density relative to that for the total body. We then write

$$3\bar{\rho}=y^n \tag{3.31b}$$

where the index n is given by

$$B=1+1/n. \tag{3.31c}$$

It is appropriate to introduce the new distance variable x according to

$$x=(\theta/n)^{1/2}r. \tag{3.31d}$$

Inserting the various expressions (3.31) into equation (3.29) for the mass transforms the mass equation into the density equation:

$$\frac{\mathrm{d}^2y}{\mathrm{d}x^2}+\frac{2}{x}\frac{\mathrm{d}y}{\mathrm{d}x}+y^n=\phi_B\left(\frac{n}{\theta_B}\right)^{(2n+1)/2n}x^{3(n+1)/n}y^{(n-1)/n}. \tag{3.32}$$

This is a non-linear equation, like equation (3.29), except when $n=1$, corresponding to $B=2$ when it becomes linear (see equation (3.31c)). For a non-rotating body, equation (3.32) reduces to

$$\frac{d^2y}{dx^2}+\frac{2}{x}\frac{dy}{dx}+y^n=0 \tag{3.33}$$

and this form is usually called the Emden equation. For planetary problems, n lies in the range $0<n<1$ and often $B\sim 3$ is a useful approximation in practice.

The boundary conditions again must be based on two criteria because we are dealing with a second-order equation. One will be the density at the surface. This will be that of silicate material for a terrestrial-type planetary body ($\rho \gtrsim 2500$ kg m^{-3}); for a fluid planet it will be the density of the liquid/dense vapour interface. The second could be the gradient of the density at the surface region, but this might well be unknown initially. The density at the centre would be a useful boundary condition from a mathematical point of view, but is unknown before the calculation is made. It can be assigned provisionally for a particular calculation and subsequently adjusted in later calculations to provide the observed total mass using equation (3.11a). Obtaining an appropriate magnitude for the inertia factor (see §3.6) can be used as a check on the usefulness of a particular calculation.

3.5.3 *EXPLICIT ACCOUNT OF THE CENTRAL DENSITY*

The central density does not appear explicitly in the deduction of equation (3.3) but it has been seen useful to rearrange our arguments to show this explicitly. The Emden equation readily allows this; we will derive the Emden equation in an alternative way, not involving equation (3.29) for the mass, which brings out the underlying assumption of hydrostatic equilibrium.

The compression of material is accompanied by an increase of temperature, the energy arising from that of the compression forces. At constant pressure, the change of energy in moving from an initial to a final state which is different is described thermodynamically as the change of enthalpy. For planetary interiors, however, the pressure changes and the energy change in moving downwards in the volume is the integral $\int_1^2 (dp/\rho)$ for a unit mass between the radius R_1 and the radius R_2. This integral, called the Bernoulli integral in fluid mechanics, can be calculated both from the elastic properties of the material and separately from the gravitational energy of the assembly of material forming the planet. Equating the two expressions gives an equation which can be rearranged to be the Emden equation for the material density.

Using the equation of state (3.20) we have

$$\int_{\rho_0}^{\rho}\frac{dp}{\rho}=\frac{K_0}{\rho_0^B}\int_{\rho_0}^{\rho}\rho^{(B-2)}\,d\rho$$

$$=\frac{nK_0}{\rho_0}\left[\left(\frac{\rho}{\rho_0}\right)^{1/n}-1\right] \tag{3.34}$$

where $n=1/(B-1)$ according to (3.31c).

Alternatively, the condition of hydrostatic equilibrium is applied using the modification

$$\int_0^p \frac{\mathrm{d}p}{\rho} = \int_0^R \frac{1}{\rho}\frac{\mathrm{d}p}{\mathrm{d}r}\,\mathrm{d}r.$$

Using equation (3.14) this gives alternatively

$$\int_0^R \frac{\mathrm{d}p}{\rho} = -\int_{R_P}^R \left(\frac{GM(r)}{r^2} - \tfrac{2}{3}w^2 r\right)\mathrm{d}r$$

$$= -G\int_{R_P}^R \frac{M(r)}{r^2}\,\mathrm{d}r + \tfrac{1}{3}w^2(R^2 - R_P^2). \tag{3.35}$$

Equating (3.34) and (3.35) gives the relation

$$n\frac{K_0}{\rho_0}\left[\left(\frac{\rho}{\rho_0}\right)^{1/n} - 1\right] = G\int_R^{R_P} \frac{M(r)}{r^2}\,\mathrm{d}r + \tfrac{1}{3}w^2(R^2 - R_P^2).$$

Differentiating both sides of this equation with respect to R:

$$\frac{K_0 \rho^{1/n-1}}{\rho_0^{B+1}}\frac{\mathrm{d}p}{\mathrm{d}R} = -\frac{GM(R)}{R^2} + \tfrac{2}{3}wR^2.$$

Using equation (3.11a) and differentiating the equation a second time with respect to R gives, in the absence of rotation

$$\frac{1}{R^2}\frac{\mathrm{d}}{\mathrm{d}R}\left(R^2\frac{\mathrm{d}y}{\mathrm{d}R}\right) + \left(\frac{4\pi G\rho_0^n \rho_c}{n\rho_c^{1/n}K_0}\right)y^n = 0 \tag{3.36}$$

where ρ_c is the central density and now $y = \rho/\rho_c$.

We now introduce the new unit of length D by

$$D^2 = \frac{n\rho_c^{1/n-1}K_0}{4\pi G\rho_0^n} \tag{3.37}$$

and the dimensionless distance $x = rD$ to obtain the equation

$$\frac{1}{x^2}\frac{\mathrm{d}}{\mathrm{d}x}\left(x^2\frac{\mathrm{d}y}{\mathrm{d}x}\right) + y^n = 0. \tag{3.38}$$

This is the Emden equation. The natural boundary conditions are seen to be $y(0) = 1$ and $\mathrm{d}y(0)/\mathrm{d}x = 0$ at the centre, the central gradient of density being zero. These conditions are straightforward to apply and the utility of introducing the central density is clear for numerical computation, even though it is strictly an unknown of the problem.

3.6 THE INERTIA FACTOR

The moment of inertia I_0 of the planetary body about the rotation axis is

$$I_0 = \frac{8\pi}{3} \int_0^{R_0} \rho(r) r^4 \, \mathrm{d}r \tag{3.39}$$

where R_0 now refers to the radius of the sphere of volume equal to that of the spheroidal shape the rotating planet actually assumes. The dimensionless inertia factor α_0 is given by

$$\alpha_0 = I_0 / M_0 R_0^2 = \frac{8\pi}{3} \int_0^1 \rho(r) r^4 \, \mathrm{d}r. \tag{3.40}$$

Using equation (3.31a) this can be written alternatively

$$\alpha_P = \frac{8\pi\hat{\rho}}{9} \int_0^1 \left(\frac{\mathrm{d}m}{\mathrm{d}r}\right) r^2 \, \mathrm{d}r. \tag{3.41}$$

These expressions are to be used with the equations for the mass and density for deducing the moment of inertia of the body. They assume the material to be under sufficient pressure that conditions of hydrostatic equilibrium apply. This excludes the crust encasing the body and a comparison with observational data can be made only when the properties of the crust are also taken into account. The observed moment of inertia I_P is related to I_0 by $I_P = I_0 + I_c$, where I_c is the moment of inertia of the crust. For a silicate/ferrous body the density of the crust can usually be taken to be 2500 kg m^{-3}; for a fluid body the density is that of the vapour which can usually be neglected in comparison with that of the main body. The effect of the crust is never critical and the interior conditions are largely independent of the crustal features.

3.7 CONCLUSIONS

1 Conditions in the main volume of a planetary body are those of hydrostatic equilibrium.

2 The shape of such a body in rotation is expressed as the relative difference between the equatorial and polar radii.

3 An immediate consequence of hydrostatic equilibrium is the differentiation of material of differing densities. In terms of the cosmic abundance, the ferrous materials will sink to the central regions, the silicate materials will float above to form the central region, while the lightest materials will form a crustal/atmospheric layer on the outside.

4 The differentiation process can act quickly (by geophysical standards). For instance, a ferrous core in a terrestrial-type planetary body could accumulate in as short a time as 10^8 years.

5 The interior conditions of the body are described by a second-order non-linear differential equation for the distribution of mass, or equivalently the density.

6 The application of these equations to particular bodies requires a knowledge of the equation of state of the constituent materials. Because a planet is a cold body, this involves the dependence of density on pressure with the temperature not included.

7 Empirical equations of this type can be derived by supposing the bulk modulus of the material to be expressible as a power series in the pressure. The resulting equations of state have had considerable success in describing the behaviour of real matter.

8 The more correct expressions of state, based on statistical mechanical considerations, have so far proved both more complicated and less generally effective than the empirical formulae.

REFERENCES AND COMMENTS

RC3.1 HYDROSTATIC CONDITIONS

The laws of hydrostatics are a limiting case of the more general laws of fluid flow. These matters are considered in a variety of books: a general reference useful in geophysics is

Tritton D J 1977 *Physical Fluid Dynamics* (Victoria: Van Nostrand Reinhold)

The shape of a rotating fluid body was considered in detail by Newton in his *Principia Mathematica.* He supposed two columns of water to be set up in the Earth, one along the axis of rotation joining the centre of the Earth to the North pole and the other at right angles joining the centre to the equator. The two columns were joined at the centre and water could pass freely from one to the other (see figure 3.1). Due to the rotation of the Earth the outward force of rotation reduces the inward force of gravity on the equatorial column in comparison with the vertical (polar) column. Equilibrium between the two columns is possible only if the pressure at the centre of each is the same, because they are joined. This requires a higher head of water in the equatorial column than in the polar column, showing the equatorial radius to be larger than the polar radius. This shows the planetary figure to be that of an oblate spheroid.

There are limits to the shape and speed that a stable rotating fluid may assume. One account of these matters of great interest is

Lyttleton R A 1953 *The Stability of Rotating Liquid Masses* (London: Cambridge University Press)

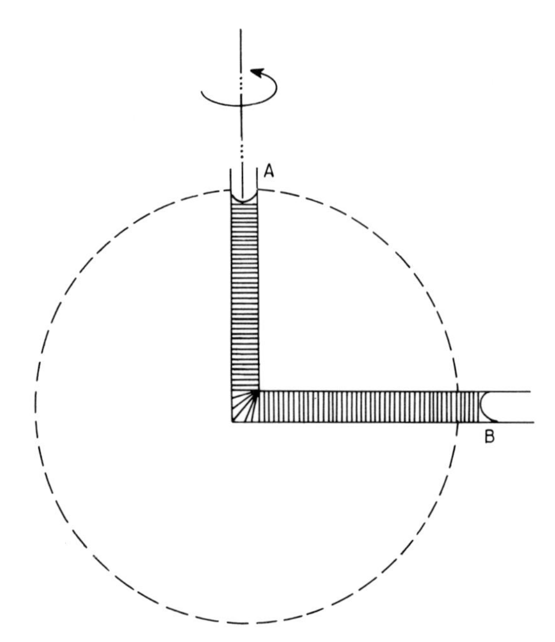

Figure 3.1 *The balance of two hypothetical columns of water along the polar and equatorial axes of a rotating body.*

RC3.2 DIFFERENTIATION OF MATERIALS

The expectation that the internal material will be differentiated is very important. Certainly this is the case in those bodies (principally the Earth and Moon) where definite information is available, and also applies in those (for instance Jupiter and Saturn) where only indirect inference can be used. It is generally supposed that the larger bodies are differentiated but the internal structure of some of the icy satellites of Saturn (see chapter 9) could be a different story. The mixture of ice and silicate material which is almost certainly their composition may well not be differentiated, the whole forming a 'dirty' ice ball.

RC3.3 INCREASE OF DENSITY WITH DEPTH

The condition of hydrostatic equilibrium implies that the density within the body will not decrease with depth as a general rule; the effects of compression will cause the density in fact always to increase except possibly for certain short-lived (geologically speaking) local variations. This was recognised by Adams and Williamson who used this condition as a starting point for determining the distribution of density within the Earth. The early paper is

Williamson E D and Adams L H 1923 Density Distribution in the Earth *J. Wash. Acad. Sci.* **13** 413–28

The method has been used and developed during the intervening years and has provided, in collaboration with seismic data for the ratio of bulk modulus to density locally, good data for the density and pressure within the Earth. The variation of the elastic constants with depth has also been determined. The inferred internal structure for the Earth is displayed in figure 8.3, p 171. Details of this work are well summarised in

Jeffreys Sir Harold 1962 *The Earth* 4th edn (London: Cambridge University Press)

This is the classic book in the field and is a mine of information and stimulation. Much work in this area was done by Bullen and his students and this is very well summarised in

Bullen K E 1975 *The Earth's Density* (London: Chapman and Hall)

This book contains many references and is very stimulating to read.

With so much data available it is interesting to attempt to determine in detail the chemical composition of the Earth as a representative terrestrial planet. Errors of observation (and there are many) are critical here. It is certainly important to determine the precise information available and particularly the consistency of the data within the context of observational uncertainty. This implies the use of sampling techniques and the matter was investigated in a very stimulating way by

Press F 1968 Earth models obtained by Monte Carlo inversion *J. Geophys. Res.* **73** 5223–34

—— 1970a Earth models consistent with geophysical data *Phys. Earth Planet. Interiors* **3** 3–22

—— 1970b Regionalised Earth models *J. Geophys. Res.* **75** 6575–8

RC3.4 EMPIRICAL EQUATIONS OF STATE

The idea of expanding the compressibility in powers of the pressure is an obvious one and it is gratifying that it has been so useful over past years. The more correct theory would be based on the arguments of the statistical mechanics of solid bodies. Here it is necessary to determine the sum over states for the system and calculate the dependence on the volume to determine the pressure and so the equation of state. To do this it is necessary first to specify the interatomic force characteristics and here there is difficulty for all but the simplest atoms. The full theory has not progressed very far in practical quantitative terms: see

Tolman R C 1938 *Statistical Mechanics* (London: Oxford University Press)

RC3.4.1 Finite strain. The theory developed in the main text assumes the equation of state for the material is the Murnaghan equation (3.20) of the main text. This is an approximate equation and a more realistic account could be expected to follow from a more detailed study of the conditions of strain in the material.

The elementary theory of elasticity assumes a linear relation between stress and strain in the material (Hooke's law) and this is closely followed by real materials providing the strain is effectively infinitesimal. Conditions deep in a planetary body might well involve finite strain and the theory must be extended to take this situation into account explicitly. There is no single way in which this can be done; see for example

Knopoff L 1964 *Q. Rev. Geophys.* **2** 625–60

One extension of the theory to include finite strain is that of

Murnaghan F D 1951 *Finite Deformation of an Elastic Solid* (New York: Wiley)

This was adapted to apply to geophysical problems by

Birch F 1952 Elasticity and the constitution of the Earth's interior *J. Geophys. Res.* **57** 227–86

The method involves the arrangement of the Helmholtz free energy A of the material to show explicitly the effect of finite strain. There is a detailed account in

Bullen K E 1975 *The Earth's Density* (London: Chapman and Hall) pp 99–107

See also

Stacey F D 1970 *Physics of the Earth* 2nd edn (New York: Wiley)
Stacey F D, Brennan B J and Irvine R D 1981 Finite strain theories and comparisons with seismological data *Geophys. Surv.* **4** 189–232.

Suppose x_i and $x_i + dx_i$ are the positions of two neighbouring points in the elastic material and suppose u_i and $u_i + du_i$ are the respective displacements due to the action of stresses. The components of the infinitesimal stress are

$$2e_{ij} = [(\partial u_i/\partial x_j) + (\partial u_j/\partial x_i)]$$

where the indices i and j take values 1, 2, 3 corresponding to the three spatial axes.

For finite strain the corresponding components are

$$2\varepsilon_{ij} = 2e_{ij} - \sum_{m=1}^{3} [(\partial u_m/\partial x_i)(\partial u_m/\partial x_j)].$$

The dilatation θ is

$$\theta = \text{div}\, \boldsymbol{u}$$

so that

$$3\varepsilon = \theta - \theta^2/6.$$

For the density ρ relative to that at zero pressure ρ_0, denoted by y, we have

$$y = \rho/\rho_0 = V_0/V = (1 - \partial u_1/\partial x_1)^3$$
$$= (1 - 2\varepsilon)^{1/2}.$$

Introduce f, called the compression, according to $f = -\varepsilon$. Then

$$y = (1 + 2f)^{3/2} \qquad \text{(RC3.1)}$$

or

$$2f = (y^{2/3} - 1).$$

The introduction of the compression is the new feature of the theory.

To obtain the equation of state of the material we invoke the arguments of statistical mechanics. Accordingly, the pressure p is the volume dependence of the free energy:

$$p = -\partial A/\partial V \qquad \text{(at constant temperature).} \qquad \text{(RC3.2)}$$

Suppose the free energy A to be related to that at zero pressure A_0 by a simple expansion in powers of f:

$$A = A_0 + \sum_{i=1}^{\infty} a_i f^i \qquad \text{(RC3.3)}$$

where the a_i are coefficients to be determined. We take A_0 to be the datum state and so set $A_0 = 0$. For reasons of thermodynamic stability there is no linear dependence on f so the expansion (RC3.3) begins with the quadratic term.

From equations (RC3.2) and (RC3.3) the pressure is written

$$p = \tfrac{3}{2} K_0 (1 + 2f)^{5/2} 2f (1 - 2a_2 f + 3a_3 f^2 + \cdots)$$

and the coefficients a_i depend generally on both the temperature and the chemical structure of the material.

The bulk modulus for the material is obtained by differentiation according to $K = \rho\, \partial p/\partial \rho$, so

$$K(p) = K_0 (1 + 2f)^{5/2} (1 + 7f - 4a_1 f)$$

if only the first power of f is retained. The dependence of K on pressure is, at constant temperature

$$\partial K(p)/\partial p = (12 + 49f - 4a_1)/3(1 + 7f).$$

In the limit of vanishing deformation ($f \to 0$) this dependence is denoted by B in the main test. Then

$$B = (12 - 4a_1)/3$$
$$= 4 - 4a_1/3.$$

This is inverted to give

$$a_1 = 3 - 3B/4. \qquad \text{(RC3.4)}$$

Many materials of planetary interest have $B \sim 4$ so a_1 is numerically very small. From equations (RC3.3) and (RC3.4) we find

$$p = \tfrac{3}{2}K_0(y^{7/3} - y^{5/3})[1 - (3 - 3B/4)(y^{2/3} - 1)]$$

as the expression for the pressure. The expression for the bulk modulus is

$$K(p) = K_0 y^{5/3}[1 + \tfrac{3}{2}(B - \tfrac{5}{3})(y^{2/3} - 1)].$$

These expressions are the Murnaghan equations of state. For materials where $B = 4$ the expression for the pressure simplifies to some extent, as does that for the bulk modulus.

At high pressures $\partial K/\partial p \to \tfrac{7}{3}$ while at low pressures $\partial K/\partial p \to 2.6$. Apparently the change of the bulk modulus with pressure is smaller at high pressures than at low, which is understandable on physical grounds.

The extended equations of Murnaghan are considerably more complicated to apply in calculations than those included in the main text (for instance equation (3.20)). At the present stage the conclusions which follow from equation (3.20) seem adequate for dealing with planetary interiors and the extended theory accounting for finite strain is a complication that most authors have been able to avoid.

RC3.5 *MASS AND DENSITY FOR A MURNAGHAN MATERIAL*

The equation for the mass is considered by

Cole G H A 1971 On Inferring Elastic Properties of the Deep Lunar Interior *Planet. Space Sci.* **19** 929–47

—— 1972 On Inferring Elastic Properties of Deep Planetary Interiors: Moon and Mars *Planet. Space Sci.* **20** 557–69

The Emden equation was applied to the investigation of the interior of Mars particularly by

Lyttleton R A 1965 On the internal structure of Mars *Mon. Not. R. Astron. Soc.* **129** 21–39

The direct deduction of the Emden equation from the principles of fluid mechanics was treated by

de Marcus W C 1958 Planetary Interiors *Encyclopaedia of Physics* (Berlin: Springer) pp 419–48

4

OBSERVED SHAPE AND INTERNAL MASS DISTRIBUTION

It was seen in §3.1 that the shape taken up by a self-gravitating planetary body in rotation shows full symmetry about the rotation axis given sufficient time to assume the equilibrium shape. The observed shape may show deviations from the ideal because the outer crustal region is not under sufficient pressure to flow, but it too can be expected to show a broadly symmetric shape. The interior conditions would be unaltered in principle even if the crust were removed, so the crust can be forgotten to a large extent when we consider the overall shape of the body. The shape taken up in fact depends on the balance between the inward acting force of gravity and the outward acting force of rotation. On the (hypothetical fluid) surface these forces balance, the resultant force being zero. This means the surface is characterised by a constant potential of the resultant force and this condition is used to calculate the shape of the free surface. The vagaries of the crust are ignored. It is necessary to know both the potential of the gravitational force and the rotation force at the surface if the shape of the free surface is to be calculated.

The strength of the rotation force is proportional to the cosine of the latitude which has its maximum at the equator and vanishes at the poles. The variation is symmetric about the equator, the corresponding surface deviation from the mean spherical shape also showing this symmetry. In these cases, the cross sectional shape is fully specified by the ellipticity e, defined by equation (3.1), which is essentially the difference between the equatorial and polar radii. This is never negative but is zero for a non-rotating mass where the equatorial and polar axes are of the same length. This is written in general $e \geqslant 0$.

The observed shape profiles, often called the figures, of the planetary bodies of the Solar System approximate very closely to this simple symmetry. The major planets show north/south symmetry, although the rapid rotation causes substantial polar flattening in these cases. The observed ellipticities are listed in table 4.1. The terrestrial planets show very much less flattening, with Venus and Mercury showing essentially none at all. When the figures are reviewed in fine detail, deviations from north/south symmetry are noticeable quite separately from hills and valleys. Not only are the broader surface details dependent on the latitude but longitude effects are also measureable. The figures of the major planets, and particularly Jupiter and Saturn, are immediately seen to be compatible with internal hydrostatic equilibrium under rotation although this is not immediately obvious for the terrestrial planets except in broad detail.

Table 4.1 *Data of the inverse flattening $1/e$ and J_2 and J_4 referring to the gravitational fields for several planets. For Mercury and Venus, J_2 and $J_4 < 10^{-5}$ and $1/e$ is essentially indefinitely great.*

Planet	$1/e$	J_2	J_4
Earth	294.12	1.0827×10^{-3}	—
Mars	169.491	1.96×10^{-3}	—
Jupiter	15.699	1.472×10^{-2}	-6.5×10^{-4}
Saturn	9.8039	1.65×10^{-2}	-1.0×10^{-3}
Uranus	41.6667	$3\text{–}5 \times 10^{-3}$	—
Neptune	37.594	5×10^{-3}	—
Moon	—	2.0272×10^{-4}	6.276×10^{-6}

4.1 GRAVITATIONAL POTENTIAL FOR AN ISOTROPIC SPHERE

The sphere is the simplest shape for theoretical discussion and is also the shape assumed by a non-rotating body. It is also approximately the shape of a slowly rotating body. In making the calculation the continuous sphere is represented as the superposition of a large number of thin spherical shells with a common centre (the onion model). Interest is focused on a representative spherical shell of radius R and thickness $\mathrm{d}R$ with $\mathrm{d}R \ll R$. The strength of the gravitational force between the shell of mass $\mathrm{d}M$ and a separate localised small mass m depends on whether the small mass is outside or inside the shell. For m *outside* the thin shell of constant density ρ, the gravitational interaction has a form which would apply if the total mass of the shell were concentrated at the geometrical centre. Then

$$\mathrm{d}M = 4\pi R^2 \rho \, \mathrm{d}R.$$

The gravitational force dF between the shell and the exterior mass m distance r ($>R$) from the centre of the sphere is

$$dF = -(4\pi G\rho R^2 m/r^2)\, dR \qquad (r > R) \tag{4.1}$$

the negative sign referring to the attractive nature of the gravitational force.

For the mass *inside* the shell ($r < R$) the force is zero. This result is the direct consequence of the inverse square form of the gravitational force applying to the spherical shell.

In terms of the mass of the shell the expression for the force accounting for the two locational possibilities is

$$dF = -(Gm\, dM/r^2) \qquad (r > R)$$

$$dF = 0 \qquad (r < R).$$

The gravitational force is associated with a gravitational potential U according to

$$U = -\int F\, dr.$$

For the mass outside the shell the potential is

$$U = -(Gm\, dM/r) \qquad (r > R). \tag{4.2}$$

For the mass inside the shell the potential is $U = \text{constant}$. This applies everywhere within the shell including $r = R$; consequently

$$U = -(Gm\, dM/R) \qquad (r < R) \tag{4.3}$$

The behaviour of a continuous spherical mass is obtained by adding together the contributions of all the separate shells. This addition does not require the densities of the shells to be the same, and for a planetary body the density will generally increase with depth. We then get the following results for the sphere.

(i) For a point outside the sphere of mass M_P and radius R_P the mass behaves as if it were concentrated at the geometrical centre. The gravitational force F between the sphere and the small mass m outside is

$$F = -(GM_P m/r^2) \qquad (r > R_P) \tag{4.4}$$

corresponding to the potential

$$U = -(GM_P m/r). \tag{4.5}$$

(ii) For a point within the sphere, the contribution is in two parts. Those shells outside the point in question do not contribute to the force: those shells inside do, as if their masses were concentrated at the centre. For the point r within the sphere, the gravitational force on the mass m is

$$F(r) = -(GM(r)m/r^2) \qquad (r < R_P) \tag{4.6}$$

where $M(r)$ is the mass contained within the sphere of radius r. The mass in the shell contained by $r<R<R_P$ makes no contribution to the force.

For a homogeneous sphere, where the density is everywhere the same, we have

$$(M(r)/r^3)=(M_P/R_P^3)$$

and equation (4.6) becomes

$$F(r)=-(GM_Pm/R_P^3)r \tag{4.7}$$

within the spherical body.

4.2 THE EXTERNAL GRAVITATIONAL FIELD

The potential U of the gravitational interaction for a large spherical mass M_P and a small external mass m is given by equation (4.5). In plotting the gravitational field external to the mass M_P it is convenient to consider the interaction with a unit external mass ($m=1$), and this gives the potential of the external gravitational field in the form

$$U=-(GM_P/r). \tag{4.8}$$

The external field does not, according to its selection, involve a region of space containing mass. The potential is in consequence a solution of the Laplace equation

$$\nabla^2 U=0 \tag{4.9}$$

where ∇^2 is the Laplacian. For complete spherical symmetry this takes the form

$$\frac{1}{r^2}\frac{\partial}{\partial r}\left(r^2\frac{\partial U}{\partial r}\right)=0. \tag{4.10}$$

It is readily found by direct differentiation and substitution that $1/r$ is a solution of equation (4.10) and that equation (4.8) is the dimensional form appropriate to gravitation.

For full three-dimensional symmetry, with latitude θ and longitude λ, the Laplace equation becomes

$$\frac{1}{r^2}\frac{\partial}{\partial r}\left(r^2\frac{\partial U}{\partial r}\right)+\frac{1}{r^2\sin\theta}\frac{\partial}{\partial\theta}\left(\sin\theta\frac{\partial U}{\partial\theta}\right)+\frac{1}{r^2\sin^2\theta}\frac{\partial^2 U}{\partial\phi^2}=0 \tag{4.11}$$

where $U=U(r,\theta,\phi)$.

The solution of equation (4.11) appropriate to the region outside the mass is well known to be expressible as an expansion in inverse powers of r and involving the harmonic angular Legendre and associated functions $P_n(\mu)$ and

$P_n^m(\mu)$ with $\mu = \cos\theta$ and n and m integers ($m > n$). These functions will be introduced later as they are required. The expansion for U is then written as the series

$$U = -\frac{GM_P}{r}\left[1 + \sum_{n=2}^{\infty}\left(\frac{a}{r}\right)^n J_n P_n(\mu) + \sum_{n=2}^{\infty}\sum_{m=1}^{\infty}\left(\frac{a}{r}\right)^n (C_{nm}\cos m\lambda + S_{nm}\sin m\lambda)P_n^m(\mu)\right] \quad (4.12)$$

where J_n, C_{nm} and S_{nm} are numerical coefficients and a is the mean planetary radius. For symmetry about the polar axis, equation (4.12) reduces to the simpler form

$$U = -\frac{GM_P}{r}\left[1 + \sum_{n=2}^{\infty}\left(\frac{a}{r}\right)^n J_n P_n(\mu)\right]. \quad (4.13)$$

This expansion is reduced still further if the north and south hemispheres are symmetric reflections one of the other. Then only the even coefficients are retained in equation (4.13).

The application of equation (4.13) to observed planetary shapes need not involve many terms of the expansion if very high accuracy is not required. Usually $J_2 \sim 10^{-3}$ while $J_4 \sim 10^{-4}$ or less. Numerical details are collected in table 4.1 for several planets.

4.3 SHAPE OF THE EXTERNAL SURFACE

The description of the observed profile of a planetary body with symmetry about the rotation axis is made in terms of the ellipticity e defined in equation (3.1). That this is possible shows that the profile is closely spheroidal. The small magnitudes of e listed in table 4.1 imply that the elliptic cross section will provide a good zero-order approximation to the actual shape, the full body of rotation being an oblate spheroid.

The location of the point on the surface of such a body is assigned by two numbers: r, the distance from the geometrical centre and θ, the co-latitude, being the angular distance from the pole to the point on the surface. The usual latitude angle is $\pi/2 - \theta$. The length of the semi-major axis is a, while that of the semi-minor axis is $a(1-e)$, as shown in figure 4.1.

The equation to the surface is

$$r^2\left(\frac{\cos^2\theta}{a^2(1-e)^2} + \frac{\sin^2\theta}{a^2}\right) = 1. \quad (4.14)$$

Remembering that e is small, r can be expanded in powers of e, the series being terminated at a term determined by the particular degree of approximation required. For instance, for the Earth, terms involving the cube or higher

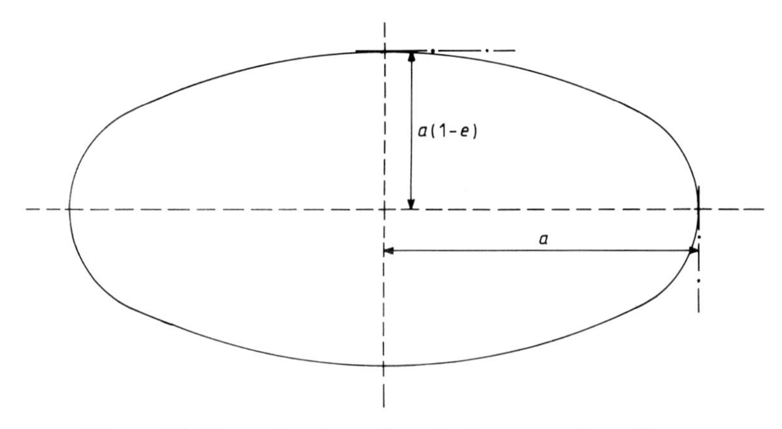

Figure 4.1 *The semi-major and semi-minor axes of an ellipse.*

powers of e can usually be neglected so that equation (4.14) becomes

$$r=a(1-e\cos^2\theta-\tfrac{3}{2}e^2\sin^2\theta\cos^2\theta). \tag{4.15}$$

The angular terms can be written in terms of the lower order zonal harmonics, given in RC4.1. In particular, writing

$$P_2(\mu)=\tfrac{1}{2}(3\cos^2\theta-1)$$

$$P_4(\mu)=\tfrac{1}{8}(35\cos^4\theta-30\cos^2\theta+3)$$

we have alternatively

$$\cos^2\theta=\tfrac{1}{3}(2P_2(\mu)+1)$$

$$\sin^2\theta\cos^2\theta=-\tfrac{8}{35}P_4(\mu)+\tfrac{2}{21}P_2(\mu)+\tfrac{2}{15}.$$

This gives the equation to the spheroidal surface (4.15) in the alternative form

$$r=a[(1-\tfrac{1}{3}e-\tfrac{1}{5}e^2)-(\tfrac{2}{3}e+\tfrac{1}{7}e^2)P_2(\mu)+\tfrac{12}{35}e^2P_4(\mu)]. \tag{4.16}$$

If we introduce the zero-order harmonic $P_0=1$ we can write equation (4.16) in the form of the series

$$r/a=A_0P_0+A_2P_2(\mu)+A_4P_4(\mu) \tag{4.17}$$

where

$$A_0=1-\tfrac{1}{3}e-\tfrac{1}{5}e^2$$

$$A_2=-(\tfrac{2}{3}e+\tfrac{1}{7}e^2)$$

$$A_4=\tfrac{12}{35}e^2.$$

The form (4.16) is to be remembered as applying to a body with symmetry about the rotation axis. It involves ascending orders of zonal harmonics and the coefficients also involve ascending powers of the ellipticity e. Only even harmonics are present (P_0 being regarded of even order) because the body is supposed to show full symmetry about the rotation axis.

The surface shape (4.16) reproduces the planetary profile *precisely* only for a homogeneous body and so is a special case of a more general shape. The addition of even harmonic functions of higher order than four will provide a surface still showing symmetry about the rotation axis but with a cross sectional profile that is not exactly elliptic. The additional terms also will involve higher-order powers of e and become the more important the faster the rotation of the body as a whole. The more general equation to the surface is written

$$r/a = A_0 + \sum_{n=1}^{\infty} A_{2n} P_{2n}(\mu). \tag{4.18}$$

Notice that $n = 1, 2, 3, \ldots$ so only even harmonics are involved.

The observed surface profile of a fluid planet (such as Jupiter or Saturn) will show full symmetry about the rotation axis and about the equatorial plane but a body with a solid crust (such as a terrestrial planet) will not show such complete symmetry in fine detail, although it could well show approximate axial symmetry. Deviations from the full symmetry are found for the Earth amounting to a few metres either side of a mean shape. Such deviations are accounted for by modifying equation (4.18) appropriately.

The measured asymmetry between the north and south hemispheres can be included by adding the odd-order harmonics to equation (4.18). The first term $P_1(\mu)$ represents a rigid body translation and so is not relevant to our discussion. Consequently, we introduce the odd harmonics starting with $P_3(\mu)$ and so extend equation (4.18) to the form

$$r/a = A_0 + \sum_{n=2}^{\infty} A_n P_n(\mu) \tag{4.19}$$

n again being an integer.

There is no account of longitude so far but this departure from symmetry about the rotation axis is included by introducing the associated harmonics $P_n^m(\mu)$, where m is an integer satisfying the condition $m < n$. These terms will give the description of the shape the same mathematical form as equation (4.12) for the full gravitational potential and this association between shape and gravity is obviously correct.

The full expression involving longitude as well as latitude will be of importance for a terrestrial-type body only for the most detailed calculations, particularly involving data from artificial satellites. These deviations from simple symmetry would be important in transient circumstances only, for instance if full-body oscillations were to occur.

4.4 EQUIPOTENTIAL SURFACE

A planetary body in rotation assumes an equilibrium shape for which the gravitational potential is everywhere the same on the surface. Such a surface for the Earth is called the geoid and is the surface which would be taken up by the surface of the oceans if they covered the entire surface. The determination of the geoid is important because it represents the true sea-level in cartography and the accurate determination of the geoid is a central problem of geodesy. It is necessary to specify the surface of constant potential for any planetary body as a preliminary to accurate plotting of surface detail.

The specification of the equipotential is complete when the ellipticity of the surface is related to the coefficients J_n, C_{nm} and S_{nm} of the gravitational potential. To achieve this, to a prescribed degree of accuracy, the equation for the surface in a form such as equation (4.19) is inserted into equation (4.12) for the potential and the condition introduced that the resulting surface is an equipotential. This means that the expression representing it must not involve the latitude (or in our terms the co-latitude) or longitude angles.

Consider a planetary body rotating with angular speed w. The figure will be determined by the combined action of the gravitational and rotation forces. If the rotation is not too rapid, the figure will be an oblate ellipsoid and the equation to the surface is equation (4.14). The gravitational potential U will be given by equation (4.13) with only the even terms retained because there is symmetry about the equatorial plane, and for small ellipticity we can terminate the series at the term involving J_4. This allows us to write

$$U = -(GM_{\mathrm{P}}/a)[1+(a/r)^3 J_2 P_2(\mu)+(a/r)^5 J_4 P_4(\mu)].$$

The potential U_{r} for the rotation is

$$U_{\mathrm{r}} = -\tfrac{1}{2}(r^2 w^2 \cos^2 \theta)$$

so the total potential $U_{\mathrm{T}} = U + U_{\mathrm{r}}$ is

$$U_{\mathrm{T}} = -\frac{GM_{\mathrm{P}}}{a}\left[1+\left(\frac{a}{r}\right)^3 J_2 P_2(\mu)+\left(\frac{a}{r}\right)^5 J_4 P_4(\mu)\right]-\frac{r^2 w^2}{2}\left(\frac{1}{3}+\frac{2}{3}P_2(\mu)\right). \tag{4.20}$$

The equation to the surface is equation (4.14) which is written in the alternative form

$$r^2 = \frac{a^2(1-e)^2}{\cos^2\theta+(1-e)^2\sin^2\theta}. \tag{4.21}$$

The ellipticity is small and we can expand equation (4.21) in powers of e using the binomial expansion. We need powers of (a/r) and also r^2 in equation (4.20) and we find from equation (4.21) that

$$
\begin{aligned}
a/r &= 1+e\cos^2\theta+\tfrac{3}{2}e^2\cos^2\theta-\tfrac{1}{2}e^2\cos^4\theta \\
(a/r)^3 &= 1+3e\cos^2\theta+\tfrac{3}{2}e^2(3\cos^2\theta+\cos^4\theta) \\
(a/r)^5 &= 1+5e\cos^2\theta+\tfrac{15}{2}e^2(\cos^2\theta-\cos^4\theta) \\
r^2 &= a^2[1+8e^2-(2e-11e^2)\cos^2\theta+4e^2\cos^4\theta]
\end{aligned}
\tag{4.22}
$$

neglecting terms of order e^3 or smaller.

Inserting the several expressions (4.22) into equation (4.20) gives the rather lengthy formula for the total potential including rotation:

$$
\begin{aligned}
U_T = -(GM_P/a)\{&1+e\cos^2\theta+\tfrac{3}{2}e^2\cos^2\theta-\tfrac{1}{2}e^2\cos^4\theta \\
&+\tfrac{1}{2}J_2(3\cos^2\theta-1)[1+3e\cos^2\theta+\tfrac{3}{2}e^2(3\cos^2\theta+\cos^4\theta)] \\
&+\tfrac{1}{8}J_4[1+5e\cos^2\theta+\tfrac{15}{2}e^2(\cos^2\theta+\cos^4\theta)(35\cos^4\theta-3\cos^2\theta+3)] \\
&+\tfrac{1}{2}\omega[1+8e^2-\cos^2\theta(2e-11e^2)+4e^2\cos^4\theta](1-\cos^2\theta)+O(e^3)\}.
\end{aligned}
\tag{4.23}
$$

Here

$$\omega=\frac{wa^3(1-e)}{GM_P} \tag{4.24}$$

and is the ratio of the centripetal acceleration at the equator $[w^2a(1-e)]$ to the acceleration of gravity there (GM_P/a^2). ω is of the same order of magnitude as e.

If equation (4.23) is to be an equipotential surface the expression must not involve the latitude angle. This means that the coefficients of the terms involving $\cos^2\theta$ and $\cos^4\theta$ must vanish identically. For the terms involving $\cos^2\theta$ in equation (4.23) we must then have

$$e+\tfrac{3}{2}e^2+\tfrac{3}{2}J_2(1-e)-\tfrac{15}{4}J_4-\tfrac{1}{2}\omega(1+3e)=0 \tag{4.25a}$$

while for the terms involving $\cos^4\theta$:

$$-\tfrac{1}{2}e^2+\tfrac{9}{2}eJ_2+\tfrac{35}{8}J_4+\omega e=0. \tag{4.25b}$$

The two conditions are solved simultaneously to provide expressions for J_2 and J_4. To first order in e, equation (4.25a) gives

$$e+\tfrac{3}{2}J_2-\tfrac{1}{2}\omega=0$$

or

$$J_2=-\tfrac{2}{3}(e-\tfrac{1}{2}\omega). \tag{4.26a}$$

Substituting equation (4.26a) into equation (4.25b) gives to second order

$$J_4=\tfrac{1}{15}(7e^2-20e\omega). \tag{4.26b}$$

Returning to equation (4.25a) we insert equation (4.26b) and retain terms to second order to obtain

$$J_2 = -\tfrac{2}{3}[e - \tfrac{1}{2}\omega - \tfrac{3}{2}e(3e - 2\omega)]. \tag{4.26c}$$

The expressions (4.26b) and (4.26c) allow the expansion (4.20) for the equipotential to be set down to include terms up to order e^2.

Higher-order terms are included by accounting for higher-order angular functions in equation (4.20) (which means higher powers of $\cos\theta$ than the fourth) and data for satellite trajectories nowadays make these higher terms of interest for the Earth. This is not yet the case for the other planets and the terms beyond J_4 for these introduce a complication unwarranted by the present accuracy of observational data. It is to be hoped that future space missions to other terrestrial planets will involve data of accuracy comparable to that now available for the Earth.

4.5 FORMULA FOR THE SURFACE ACCELERATION OF GRAVITY

The acceleration of gravity g_a at the surface of a planetary body distance a from the centre, is given by

$$g_a = \left[\left(\frac{\partial U_{\mathrm{T}}}{\partial r}\right)^2 + \left(\frac{1}{r}\frac{\partial U_{\mathrm{T}}}{\partial \theta}\right)^2\right]^{1/2} \tag{4.27}$$

where θ is the co-latitude. In practice, it is conventional to use the latitude angle ϕ (zero at the equator) rather than the co-latitude (zero at the pole) where $\phi = \pi/2 - \theta$. With this change we find to first order in the ellipticity

$$\frac{1}{r}\frac{\partial U_{\mathrm{T}}}{\partial \phi} = -\frac{GM_{\mathrm{P}}}{a^2}\, e \sin 2\phi$$

and $(\partial U_{\mathrm{T}}/\partial r)$ is obtained from equation (4.23). The result is

$$g_a = \frac{GM_{\mathrm{P}}}{a^2}\left[1 - \frac{2\omega}{3} - \frac{1}{3}\left(\frac{5\omega}{2} - e\right)(1 - \sin^2\phi)\right] + \mathrm{O}(e^2)$$

where we have replaced θ by ϕ in the spherical harmonic.

At the equator (where $\phi = 0$ and so $\sin\phi = 0$) we find

$$g_{\mathrm{e}} = (GM_{\mathrm{P}}/a^2)(1 + \tfrac{1}{3}e - \tfrac{3}{2}\omega)$$

so that

$$g_a = g_{\mathrm{e}}[1 + (5\omega/2 - e)\sin^2\phi] \tag{4.28}$$

over the spheroidal surface to first order in the ellipticity e. This formula for the acceleration of gravity in terms of the magnitude at the equator was first derived by Clairaut (1743).

To second order in e it is found that

$$g_a = g_{\mathrm{e}}[1 + (\tfrac{5}{2}\omega - e + \tfrac{15}{4}\omega - \tfrac{17}{14}\omega e)\sin^2\phi + e(\tfrac{15}{8}\omega - \tfrac{7}{8}e)\sin^2 2\phi].$$

The theory takes its simplest form at the latitude ϕ for which $P_2(\sin \phi)=0$, that is $\phi=\sin^{-1} \frac{1}{3}$ or $\phi=35.3°$. Then, to first order in e

$$g_a=(GM_P/a^2)(1-\tfrac{2}{3}\omega).$$

Measurements of w, e, a and g_a have led in the past to data for GM_P for the Earth and this was one of the earliest uses of the Clairaut formula.

4.6 CONDITIONS INSIDE A SPHERICAL BODY

Up to this point we have considered only the surface of the planetary body, but our main concern is with conditions inside. This introduces additional complications into the theory and it is convenient to take our arguments in several stages. In developing these arguments, we will build the analysis up in stages from the simplest case of a sphere of constant density to a non-homogeneous sphere with density increasing with depth.

4.6.1 *SPHERE OF CONSTANT DENSITY* ρ

The potential U_0 will depend upon whether we are concerned with a point r inside or outside the sphere. If R_P is the radius of the sphere and r the distance of P from the centre of the sphere then

$$U_0=(4\pi G\rho R_P^3/3r) \qquad \text{for } r \geqslant R_P \text{ (outside)}$$

$$U_0=\tfrac{2}{3}\pi G\rho(3R_P^2-r^2) \qquad \text{for } r \leqslant R_P \text{ (inside).} \tag{4.29}$$

The potential is continuous across the boundary surface of the body; these formulae satisfy this requirement as is realised by setting $r=R_P$ when both expressions become the same.

4.6.2 *ELLIPSOID OF CONSTANT DENSITY* ρ

For any point P' with co-latitude θ on an ellipsoidal surface the potential can be regarded as composed of a spherical mass of radius R_P together with a thin layer of material between the actual surface S and the associated spherical surface of radius R_P. Suppose the actual surface S to have the equation

$$r'=R_P[1+eP_2(\mu)]$$

to the first order of the ellipticity. The radial thickness of the 'extra' material at the point P, where the potential is to be found, is $eR_PP_2(\mu)$ and the element of volume is $eR_P^3P_2(\mu')\,\mathrm{d}\mu'\,\mathrm{d}\phi'$. The potential U_1 is then

$$U_1=-\int \frac{G\rho}{(PP')}\,\mathrm{d}V$$

where $\mathrm{d}V$ is the element of volume and

$$\frac{1}{PP'}=\frac{1}{R_P}\sum_{l=0}^{\infty}(R_P/r)^{l+1}P_l(\mu) \qquad (r>R_P) \text{ (outside)}$$

$$\frac{1}{PP'}=\frac{1}{R_P}\sum_{l=0}^{\infty}(r/R_P)^{l}P_l(\mu) \qquad (r<R_P) \text{ (inside).}$$

We can then write the two alternative expressions

$$U_1=-\iint\frac{G\rho}{R_P}\sum_{l=0}^{\infty}\left(\frac{R_P}{r}\right)^{l+1}P_l(\mu)eR_P^3P_2(\mu')\,\mathrm{d}\mu'\,\mathrm{d}\phi' \qquad (r>R_P)$$

$$U_1=-\iint\frac{G\rho}{R_P}\sum_{l=0}^{\infty}\left(\frac{r}{R_P}\right)^{l+1}P_l(\mu)eR_P^3P_2(\mu')\,\mathrm{d}\mu'\,\mathrm{d}\phi' \qquad (r<R_P).$$

According to the orthogonality condition for surface harmonic functions

$$\iint P_2(\mu')P_l(\mu)\,\mathrm{d}\mu'\,\mathrm{d}\phi'=\frac{8\pi}{15}P_2(\mu) \qquad \text{if } l=2$$

$$=0 \qquad \text{if } l\neq 2$$

so that the twin expressions for the potential become

$$U_1=-\frac{8\pi}{15}G\rho e\frac{R_P^5}{r^3}P_2(\mu) \qquad (r>R_P)$$

$$U_1=-\frac{8\pi}{15}G\rho er^2P_2(\mu) \qquad (r<R_P). \qquad (4.30)$$

It is customary in the subject to use a modified definition $p_2(\mu)$ for the harmonic and we will explain this now. We introduce the new second-order harmonic $p_2(\mu)$ by

$$p_2(\mu)=-\tfrac{2}{3}P_2(\mu)$$

and at the same time introduce the latitude angle ϕ in place of the co-latitude angle θ. Then

$$p_2(\phi)=\tfrac{1}{3}-\sin^2\phi. \qquad (4.31)$$

The total potential U_P at the point P is the sum of the two corresponding expressions (4.29) and (4.30);

$$U_P=U_0+U_1$$

so that, using equation (4.31) we have

$$U_P=4\pi G\rho\left(\frac{R_P^3}{3r}+\frac{0.2e}{r^3}R_P^5p_2(\phi)\right) \qquad (r>R_P) \qquad (4.32a)$$

$$U_P=4\pi G\rho[\tfrac{1}{6}(3R_P^2-r^2)+0.2r^2ep_2(\phi)] \qquad (r<R_P). \qquad (4.32b)$$

For a homogeneous ellipsoidal body with constant density throughout, the gravitational potential at a point outside is given by equation (4.32a) while for a point inside it is given by equation (4.32b).

4.6.3 THIN ELLIPSOIDAL SHELL OF CONSTANT DENSITY

For a thin shell of constant density the formulae (4.32) apply but now the spheroidal profile will differ between the two shells. The inner surface will have the profile

$$r_i = q[1 + ep_2(\mu)]$$

while the outer one will have the different profile

$$r_o = (q + \mathrm{d}q)[1 + (e + \mathrm{d}e)p_2(\mu)]$$

where $\mathrm{d}q$ is the elementary thickness of the shell.

The potentials $\mathrm{d}U_{\mathrm{Po}}$ outside the shell or $\mathrm{d}U_{\mathrm{Pi}}$ inside the shell are obtained by differentiating the expressions (4.32) with respect to q remembering that the ellipticity e also varies with q. Then we find

$$\mathrm{d}U_{\mathrm{Po}} = 4\pi G\rho[(q^2/r)\,\mathrm{d}q + \tfrac{1}{5}(1/r^3)p_2(\mu)\,\mathrm{d}(eq^5)] \tag{4.33a}$$

$$\mathrm{d}U_{\mathrm{Pi}} = 4\pi G\rho[q\,\mathrm{d}q + \tfrac{1}{5}r^2 p_2(\mu)\,\mathrm{d}e]. \tag{4.33b}$$

Again ρ is constant.

4.6.4 SPHEROID OF VARIABLE DENSITY

The variation of density within an inhomogeneous body is accounted for by supposing it to be composed of a collection of many thin spheroidal shells. The density is constant within each shell but varies in some prescribed way from one shell to the next. The effect of the total body at a chosen point is then obtained in the usual way by adding together the contributions of all the shells.

For a point external to all the shells we integrate equation (4.33a) over the full radius a of the collection of shells. This gives

$$U_{\mathrm{Po}} = \frac{G}{r}\int_0^a 4\pi\rho q^2\,\mathrm{d}q + \frac{4\pi G}{5r^3}\,p_2(\mu)\int_0^a \rho\,\frac{\mathrm{d}(eq^5)}{\mathrm{d}q}\,\mathrm{d}q$$

remembering that ρ is now a function of q and that

$$\mathrm{d}(eq^5) = \frac{\mathrm{d}(eq^5)}{\mathrm{d}q}\,\mathrm{d}q.$$

The integral forming the first term on the right-hand side is the expression for the total mass M_{P}. The integral in the second term cannot be interpreted so directly and is best left alone for the present. This means that we have

$$U_{\mathrm{Po}}=\frac{GM_{\mathrm{P}}}{r}\left[1+\frac{3}{2}K_2\left(\frac{a}{r}\right)^2 p_2(\mu)\right] \tag{4.34a}$$

where

$$K_2=\frac{8\pi}{15M_{\mathrm{P}}a^2}\int_0^a \rho(q)\frac{\mathrm{d}(eq^5)}{\mathrm{d}q}\,\mathrm{d}q. \tag{4.34b}$$

For a point P within the spheroidal volume we must account separately for the material closer to the centre than P (for which P is external) and that outside the location of P.

Suppose P to lie on the internal surface

$$r=r_1[1+e_1 p_2(\mu)].$$

Then from equations (4.33a) and (4.33b) we find

$$\begin{aligned}U_{\mathrm{Pi}}=&\frac{4\pi G}{r}\left(\int_0^{r_1}\rho(q)q^2\,\mathrm{d}q+\frac{1}{5}\frac{p_2(\mu)}{r^2}\int_0^{r_1}\rho(q)\frac{\mathrm{d}(eq^5)}{\mathrm{d}q}\,\mathrm{d}q\right)\\&+4\pi G\left(\int_{r_1}^{a}\rho(a)q\,\mathrm{d}q+\frac{1}{5}r^2p_2(\mu)\int_{r_1}^{a}\rho(q)\frac{\mathrm{d}e}{\mathrm{d}q}\,\mathrm{d}q\right)\end{aligned} \tag{4.35}$$

since $\mathrm{d}e=(\mathrm{d}e/\mathrm{d}q)\,\mathrm{d}q$.

Although it has taken rather a long argument to derive equations (4.34) and (4.35) it is necessary to do this in detail because the formulae provide the basis for investigating the interior conditions of a planet, and especially for estimating the moment of inertia about the rotation axis, as we shall see shortly.

The theory has been developed here to first order in the ellipticity e, terms involving higher powers of e being neglected. By retaining such higher-order terms in e the accuracy of the theory is improved but nothing of principle is involved in extending the theory in these terms. We therefore do not do this here and the reader who is interested in the extension is referred to the literature. For the terrestrial planets it is usually sufficient to retain terms only as far as e^2 and these developments provide the classic theories of Callendreau and of G H Darwin.

4.7 CLAIRAUT'S EQUATION FOR THE INTERNAL ELLIPTICITY

For any internal surface of constant density associated with a particular ellipticity e the associated potential U_{Pi} is a constant. The condition inside is therefore analogous to the surface conditions studied in §4.6.3. Taking account of the rotation through the potential $U_{\mathrm{R}}=\frac{1}{2}w^2r^2\sin^2\theta$ we seek the equipotential surface Ψ such that

$$\Psi = U_{\text{Pi}} + \tfrac{1}{2}w^2r^2 \sin^2 \theta = \text{constant} \tag{4.36}$$

corresponding to the inner surface profile

$$r = r_1[1 + e_1 p_2(\mu)]. \tag{4.37}$$

Introducing equation (4.35) into equation (4.36) gives the condition

$$\left(\frac{\Psi}{4\pi G}\right) = \frac{1}{r}\int_0^{r_1} q^2\rho \, \mathrm{d}q + \int_{r_1}^{a} \rho q \, \mathrm{d}q + \frac{1}{5}Lp_2(\mu) + \frac{1}{2}\left(\frac{w^2r^2 \sin^2\theta}{4\pi G}\right) = \text{constant} \tag{4.38}$$

where

$$L = \frac{1}{r^3}\int_0^{r_1} \rho(q)\frac{\mathrm{d}(eq^5)}{\mathrm{d}q}\mathrm{d}q + r^2\int_{r_1}^{a} \rho\frac{\mathrm{d}e}{\mathrm{d}q}\mathrm{d}q.$$

We will restrict the theory to first order in the ellipticity e: because e is therefore assumed to be of the first order in small quantities, we can write

$$1/r = (1/r_1)[1 - e_1 p_2(\mu)]$$
$$1/r^2 = (1/r_1^2)[1 - 2e_1 p_2(\mu)]$$
$$r^2 = r_1^2[1 + 2e_1 p_2(\mu)].$$

This arrangement allows us to separate equation (4.38) into a part proportional to $p_2(\mu)$ and a part which does not involve the latitude angle at all. This means

$$\left(\frac{\Psi}{4\pi G}\right) = L_0 + L_2 p_2(\mu) \tag{4.39}$$

where

$$L_0 = \frac{1}{r_1}\int_0^{r_1} \rho q^2 \, \mathrm{d}q + \int_{r_1}^{a} \rho q \, \mathrm{d}q + \frac{1}{3}\left(\frac{w^2r_1^2}{4\pi G}\right)$$
$$L_2 = \tfrac{1}{5}L + L_1$$
$$L_1 = -\frac{e_1}{r_1}\int_0^{r_1} \rho q^2 \, \mathrm{d}q + \frac{1}{2}\left(\frac{w^2r_1^2}{4\pi G}\right).$$

We must remember now that Ψ describes an equipotential surface and so, for given r_1, must be independent of the latitude angle. Consequently there can be no angular dependence of equation (4.39) and the coefficient L_2 must vanish. Explicitly, using equation (4.38), this condition is

$$-\frac{e_1}{r_1}\int_0^{r_1} \rho q^2 \, \mathrm{d}q + \frac{1}{5r_1^3}\int_0^{r_1} \rho\frac{\mathrm{d}(eq^5)}{\mathrm{d}q}\mathrm{d}q + \frac{r_1^2}{5}\int_{r_1}^{a} \rho\frac{\mathrm{d}e}{\mathrm{d}q}\mathrm{d}q + \frac{w^2r_1^2}{8\pi G} = 0. \tag{4.40}$$

This is the condition to be satisfied by e_1 for each value of r_1 within the ellipsoidal body. This is not the most convenient mathematical form for calculating e because it contains integrals. These can be eliminated by differentiation twice with respect to r_1. The result is a second-order differential equation for e, describing the dependence of the ellipticity on the distance from the centre:

$$\frac{\mathrm{d}^2 e_1}{\mathrm{d}r_1^2}+\left(\frac{6}{r_1}+\frac{2}{\hat{\rho}_1}\frac{\mathrm{d}\hat{\rho}_1}{\mathrm{d}r_1}\right)\frac{\mathrm{d}e_1}{\mathrm{d}r_1}+\frac{2e_1}{r_1\hat{\rho}_1}\frac{\mathrm{d}\hat{\rho}_1}{\mathrm{d}r_1}=0 \tag{4.41}$$

where $\hat{\rho}_1$ is the mean density of the material inside the spheroidal surface of mean radius r_1:

$$\hat{\rho}_1=\frac{3M_1}{4\pi r_1^3}=\frac{3}{r_1^3}\int_0^{r_1}\rho q^2\,\mathrm{d}q. \tag{4.41a}$$

It follows further that

$$\frac{\mathrm{d}\hat{\rho}_1}{\mathrm{d}r_1}=-\frac{3}{r_1}(\hat{\rho}_1-\rho_1) \tag{4.41b}$$

where ρ_1 is the density on the surface (as opposed to the mean density of the material contained within it).

This equation was first derived by Clairaut in 1743 during his studies of the interior of the Earth. There are a variety of alternative forms of the Clairaut equation (4.41), each particularly convenient for use in particular circumstances.

4.8 RADAU'S PARAMETER

Radau came across an extraordinary transformation which allows Clairaut's equation to be written in a remarkable form. Denoting the Radau parameter by η_1, where

$$\eta_1=\frac{r_1}{e_1}\frac{\mathrm{d}e_1}{\mathrm{d}r_1} \tag{4.42}$$

the Clairaut equation (4.41) can be put into the alternative form

$$\frac{\mathrm{d}}{\mathrm{d}r_1}(\hat{\rho}_1(1+\eta_1)^{1/2})+\frac{(5\eta_1+\eta_1^2)}{2r_1(1+\eta_1)^{1/2}}\hat{\rho}_1=0.$$

Introducing the function $\psi(\eta_1)$ by

$$\psi(\eta_1)=(1+\tfrac{1}{2}\eta_1-\tfrac{1}{10}\eta_1^2)(1+\eta_1)^{-1/2} \tag{4.43}$$

the Clairaut equation takes the form

$$\frac{\mathrm{d}}{\mathrm{d}r_1}(\hat{\rho}_1 r_1^5(1+\eta_1)^{1/2})=5\hat{\rho}_1 r_1^4\psi(\eta_1). \tag{4.44}$$

The importance of this equation lies in the particular behaviour of $\psi(\eta_1)$ as η_1 changes when the theory is applied to the study of the Earth.

It is found from observation for the Earth that $0 \leqslant \eta_1 \leqslant 0.6$ and the behaviour of $\psi(\eta_1)$ in this range is shown below. It is extraordinary that within this range $|\psi(\eta_1) - 1| < 8 \times 10^{-4}$ and that we can set $\psi(\eta_1) = 1$ to an accuracy better than one part in 1000, as is seen in table 4.2.

Table 4.2 *Magnitudes of the function $\psi(\eta_1)$ for a range of magnitudes of η_1.*

η_1	$\psi(\eta_1)$	η_1	$\psi(\eta_1)$
0	1.0000	0.7	0.997 821 5
0.1	1.000 182	0.8	0.995 795 6
0.2	1.000 507	0.9	0.993 177 0
0.3	1.000 723	1.0	0.989 949 5
0.4	1.000 663	2.0	0.923 760 4
0.5	1.000 208 3	3.0	0.800 000
0.6	0.999 279 7		

To this accuracy, equation (4.44) becomes

$$\frac{\mathrm{d}}{\mathrm{d}r}[\hat{\rho}_1 r_1^5 (1 + \eta_1)^{1/2}] = 5\hat{\rho}_1 r_1^4 \tag{4.45}$$

which is the approximate form taken by Clairaut's equation in these circumstances. It is important to realise that equation (4.45) is an approximation which has been found useful for the Earth although it is not necessarily applicable to other planetary bodies (for example, the major planets). The approximation for the Earth is undoubtedly good. Jeffreys has shown that its use leads to calculated data for e_a (referring to the terrestrial surface) which are too large by an error of only 6×10^{-7}: if $e_a \sim 10^{-3}$ the error is no larger than one part in 1000. At the surface of the Earth $\eta_a = 0.570$ giving $\psi(r_a) = 1.00074$.

4.9 THE INERTIA FACTOR

For a spheroidal body of mass M_P and mean radius R_P, the moment of inertia I_P about the rotation axis is given by

$$I_\mathrm{P} = \alpha_\mathrm{P} M_\mathrm{P} R_\mathrm{P}^2$$

where α_P is a number called the inertia factor. For a homogeneous sphere (of constant density throughout), $\alpha_\mathrm{P} = 0.4$; for a density distribution more concentrated towards the centre $\alpha_P < 0.4$, and the greater the central concentration the smaller is the magnitude of α_P. The arguments developed so far describe the variation of density within a planetary body and it is possible to extend them to obtain information about the inertia factor introduced in RC1.4.

The starting point is the expression of the definition for the moment of inertia about the polar axis Oz which is

$$I_P = \int \rho x^2 \, dV = 2\pi \iint \rho r^2 \sin^2\theta \, r^2 dr \sin\theta \, d\theta$$

or

$$I_P = \alpha_P M_P R_P^2 = \frac{8\pi}{3} \int_0^{R_P} \rho r^4 \, dr. \tag{4.46}$$

This expression is rearranged using the Radau equation (4.44) in its integral form

$$\hat{\rho}_1 r_1^5 (1+\eta_1)^{1/2} = 5\psi(\eta_1) \int_0^{R_P} \hat{\rho} r^4 \, dr. \tag{4.47}$$

Combining equations (4.46) and (4.47) there follows the expression for the inertia factor

$$\alpha_P = \frac{2}{3}\left(1 - \frac{2}{5}\frac{(1+\eta_a)^{1/2}}{\psi(\eta_a)}\right) \tag{4.48}$$

at the surface and a corresponding expression for any point inside with radius r_1. Equation (4.48) can be rearranged to involve ω and e_a with the result

$$\eta_P = \frac{5}{2}\frac{\omega}{e_P} - 2 \tag{4.49}$$

where

$$\omega = \frac{w^2 R_P^3}{G M_P}$$

to zero order in e_P. Inserting equation (4.49) into equation (4.48) gives

$$\alpha_P = \frac{2}{3}\left[1 - \frac{2}{5}\frac{1}{\psi(\eta_P)}\left(\frac{5}{2}\frac{\omega}{e_P} - 1\right)^{1/2}\right]$$

for conditions of hydrostatic equilibrium. Under the Radau approximation, $\psi = 1$ with the result that

$$\alpha_P = \frac{2}{3}\left[1 - \frac{2}{5}\left(\frac{5}{2}\frac{\omega}{e_P} - 1\right)^{1/2}\right]. \tag{4.50}$$

Observational data for ω and e_P allow α_P to be calculated and the results are as reliable as the assumption of hydrostatic equilibrium. This expression has been widely used in planetary studies to infer approximate data for the inertia factors of planets. It is restricted to the two conditions of hydrostatic equilibrium and a linear approximation for the figure retaining only terms of order e_P. While hydrostatic equilibrium is a satisfactory assumption within the main body of the planet it will not apply in the crust, and it is the crust that is observed.

Other alternative expressions for α_P can be constructed. Thus it is known that (ω/e_P) is related to the coefficient J_2 in the expansion of the gravitational potential in spherical harmonic functions. Explicitly, to the linear term in e_P

$$J_2=\tfrac{2}{3}e_P-\tfrac{1}{3}\omega=\tfrac{2}{3}e_P[1-\tfrac{1}{2}(\omega/e_P)].$$

Then from equation (4.49)

$$\eta_P=3[1-\tfrac{5}{2}(J_2/e_P)]$$

and from equation (4.50)

$$\alpha_P=\frac{2}{3}\left[1-\frac{4}{5}\left(1-\frac{15}{8}\frac{J_2}{e_P}\right)^{1/2}\right]$$

or alternatively

$$\alpha_P=\frac{2}{3}\left[1-\frac{2}{5}\left(\frac{4\omega-3J_2}{\omega+3J_2}\right)^{1/2}\right]. \tag{4.51}$$

4.10 THE PRECESSIONAL CONSTANT AND DYNAMICAL ELLIPTICITY

Two further results will be useful. First, it is shown in RC4.3.2 that J_2 is proportional to the difference between the polar and equatorial moments of inertia

$$J_2=(I-A)/M_PR_P^2. \tag{4.52}$$

Second, the precessional constant H, often called the dynamical ellipticity, is known to be related to the moment of inertia components I and A by

$$H=(I-A)/I. \tag{4.53}$$

Consequently

$$J_2/H=I/M_PR_P^2=\alpha_P \tag{4.54}$$

the inertia factor about the polar axis. The coefficient J_2 can be measured from satellite motions and H is obtainable astronomically so that equation (4.54) allows the inertia factor to be deduced without a direct appeal to hydrostatic theory.

For the Earth, $J_2=1.0827\times10^{-3}$ and $H=3.273\times10^{-3}$, giving $\alpha_P=0.3308$. If, further, we take $\omega=3.450\times10^{-3}$, equation (4.50) gives

$$e_P=\frac{3.450\times10^{-3}}{\frac{2}{5}\{1+[\frac{5}{2}(1-\frac{3}{2}\times0.3308)]^2\}}$$

or in numerical terms $e_P=3.3347\times10^{-3}=(1/299.88)$. A recently recommended best value is $e_P=3.352\,92\times10^{-3}=1/298.247$. The reliability of the various estimates of e_P is likely to be about one part in 300 or thereabouts. We can notice that, for the Earth, $e_P\sim\omega$; more accurately, $\omega/e_P\sim1.0289$. Also, H is numerically closely the same as e_P and ω. From equation (4.49) we also have $\eta_P=0.5649$ for the Earth at the surface corresponding to $\psi(\eta_P)=0.999\,61$.

4.11 FASTER ROTATION

The theory leading to Clairaut's equation is essentially the theory of the coefficient of $P_2(\cos\theta)$ in the expression for the radial distance of an equipotential surface from the centre of the planet. So far, we have included only terms of order e in the formulae, but this restricts the theory to bodies with sufficiently slow rotation. As the rotation increases, so the centrifugal contribution to the potential increases and the equatorial radius becomes elongated relative to the polar axis. Put another way, the ellipticity increases and there comes a point when the rotation is too fast for the analysis to be curtailed at the linear or even the quadratic term in e. The analysis must be extended to include higher powers of e still. The first steps to include higher powers of the ellipticity were taken by Sir George Darwin in a classic paper including terms to e^2. During the last thirty years the matter has been taken up again, partly because space vehicles have provided the opportunity to make measurements of vastly improved accuracy, partly because we have now achieved a better understanding of the equation of state of the material comprising the planet, and partly because developments in electronic computation now give the possibility of approaching this complicated problem in numerical terms. Because $e\propto\omega$ and $\omega\propto w^2$, where w is the rotation speed, we can regard the representation of the internal conditions in powers of e alternatively as an expansion in powers of w^2. As the rotation speed increases, so it is necessary to include higher powers of ω and this can be achieved by including higher powers of e in the analysis.

The contribution of the rotation to the gravitational potential has the form

$$\tfrac{1}{2}w^2r^2\sin^2\theta=\tfrac{1}{3}w^2r^2[1-P_2(\mu)].$$

This shows that the $P_2(\mu)$ angular dependence has a special priority in defining the shape of the body. It is this angular dependence that is involved in the first-order theory but higher orders in e (and so in ω) also involve the higher coefficients J_4, J_6 and so on. The odd coefficients (J_3, J_5, etc) will not be involved because the rotational symmetry associated with hydrostatic conditions will exclude asymmetry with respect to the equatorial plane. The various formulae associated with this approach are complicated and will not be written down here. The interested reader is referred to the literature.

The full range of the lower-order even harmonic coefficients will in this way be associated with the gravitational field and with the surface shape of the planet in other than slow rotation. We shall need to refer to this later when we consider the shape of a large planet in rapid rotation (especially Jupiter and Saturn).

4.12 CONCLUSIONS

1. An expression for the gravitational potential for an inhomogeneous sphere is obtained by representing the sphere as a combination of thin homogeneous shells (an onion model) and so adding together the corresponding contributions for the gravitational potential.
2. The external gravitational field can be represented mathematically by an expansion in inverse powers of the distance and zonal harmonics. For a body symmetric about the rotation axis, the associated harmonics are not involved. For symmetry also about the equatorial plane only the even harmonics are involved.
3. The shape of the external surface can be expanded in the same terms as the gravitational field on the basis of hydrostatic equilibrium.
4. A planetary body in rotation assumes an equilibrium shape, the surface being a figure of constant gravitational potential. For the Earth this surface profile is called the geoid.
5. This specification leads to Clairaut's formula for the surface gravity.
6. Conditions inside a spherical body can equally be expressed in terms of the internal ellipticity and the theory taken to first order leads to an equation for the internal ellipticity first derived by Clairaut. This is rearranged into a more convenient form by using the transformation of Radau.
7. The theory leads to a relation between the inertia factor for the total body, assumed to be in internal hydrostatic equilibrium, and the ellipticity of the surface. The magnitude of the inertia factor for a particular planetary body can be found if the precessional constant is also known.
8. The theory is readily extended to include fast rotation, for instance of the type shown by the major planets.

REFERENCES AND COMMENTS

The theory of the figure of the Earth has been investigated by many workers. Especially good accounts are contained in

Jeffreys Sir Harold 1962 *The Earth* (London: Cambridge University Press)
Bullen K E 1975 *The Earth's Density* (London: Chapman and Hall)

The book by Bullen contains many theorems treated in a very understandable way.

The classic account is in the paper

Darwin G H 1900 The theory of the figure of the Earth carried to the second order of small quantities *Mon. Not. R. Astron. Soc.* **60** 82–124

Other references:

Clairaut A C 1743 *Théorie de la Figure de la Terre* (Paris)
Radau R R 1885 Sur la loi des densités a l'intérieur de la Terre *C.R. Acad. Sci., Paris* **100** 972
Callendrau O 1889 *Ann. Ob. Paris* **19** E1

RC4.1 THE EXTERNAL GRAVITATIONAL FIELD

The theory of the gravitational field has been the subject of many monographs, for instance

Ramsey A S 1952 *Newtonian Attraction* (London: Cambridge University Press)

The expansion in spherical harmonic functions is a well tried and valuable technique, providing for a spherical geometry the same possibilities of representation as the Fourier expansion does for the cartesian case. The reason for the usefulness arises because the gravitational potential at a distance r from the mass centre of a gravitational mass of finite size is proportional to $1/r$, and $1/r$ can be expanded in a power series of spherical functions. To see this, consider the gravitational potential at a point P distance r from the centre due to an element of mass at the point P' within the body distance R from the centre (see figure 4.2). We need to express $1/PP'$ in terms of r.

From trigonometry we know that

$$PP' = (r^2 - R^2 - 2rR\cos\theta)^{1/2}$$

so that

$$1/PP' = (1/r)[1 + (R/r)^2 - (2R/r)\cos\theta]^{-1/2}.$$

Provided $r > R$ this expression can be expanded in powers of (R/r) to give, as far as the quadratic term,

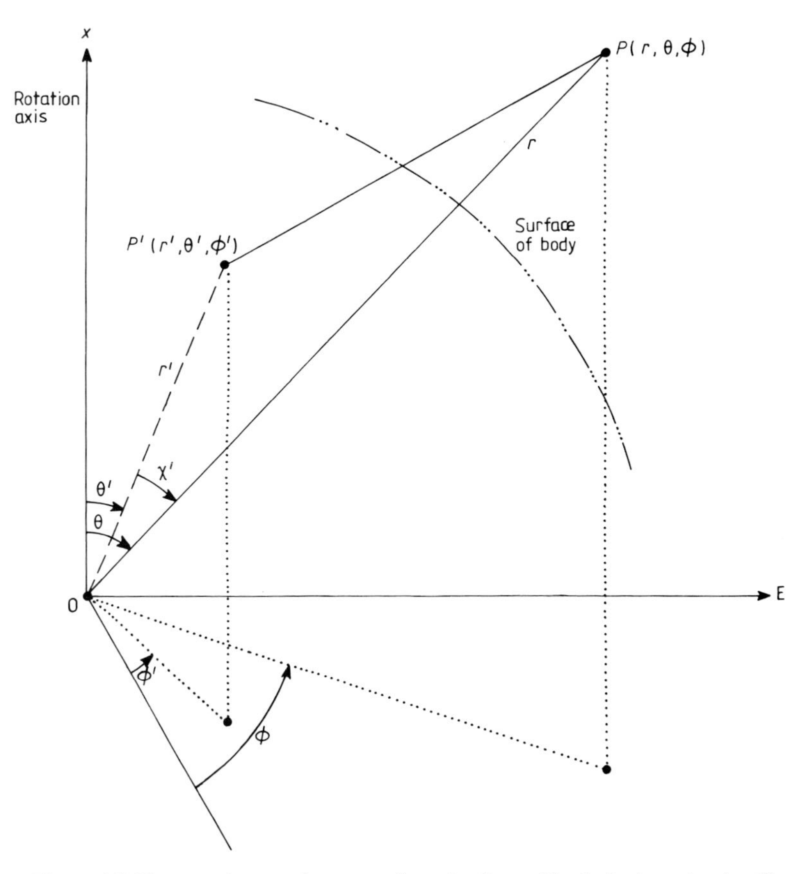

Figure 4.2 *Showing the angular array for point P outside the body and point P′ inside.*

$$1/PP' = (1/r)[1 + (R/r)\cos\theta + (R/r)^2(\tfrac{3}{2}\cos\theta^2 - 1)] + O(R/r)^3.$$

The angular terms are the zonal (spherical) harmonics and are given particular symbols.

We introduce the symbol P_n, where n is an integer, so that

$$P_1(\cos\theta) = \cos\theta$$

$$P_2(\cos\theta) = \tfrac{1}{2}(3\cos^2\theta - 1).$$

Retaining terms to the second power in (R/r) we have

$$1/PP' = (1/r)[1 + (R/r)P_1(\mu) + (R/r)P_2(\mu)]$$

where we have introduced the variable $\mu = \cos\theta$. The inclusion of higher powers of (R/r) in the expansion will introduce further angular functions into the formula. Thus

$$P_3(\mu)=\tfrac{1}{2}(5\cos^3\theta-3\cos\theta)$$

$$P_4(\mu)=\tfrac{1}{8}(35\cos^4\theta-30\cos^2\theta+3)$$

and so on. We will follow the convention of setting $\mu=\cos\theta$. The result is the expansion

$$\frac{1}{PP'}=\frac{1}{r}\left(1+\sum_{n=1}^{\infty}(R/r)^n P_n(\mu)\right)$$

for the general dependence of PP' on r.

The solution of the Laplace equation for the region outside a gravitating body introduces a wider class of angular function. For a full discussion, see for instance

Killingbeck J P and Cole G H A 1978 *Mathematical Techniques and Physical Applications* (New York: Academic)

The equation to be solved is equation (4.11) of the text, i.e.

$$\frac{1}{r^2}\frac{\partial}{\partial r}\left(r^2\frac{\partial U}{\partial r}\right)+\frac{1}{r^2\sin\theta}\frac{\partial}{\partial\theta}\left(\sin\theta\frac{\partial U}{\partial\theta}\right)+\frac{1}{r^2\sin^2\theta}\frac{\partial^2 U}{\partial\phi^2}=0. \tag{RC4.1}$$

The unknown function $U(r, \theta, \phi)$ is separated into three separate functions by

$$U(r, \theta, \phi)=R(r)V(\theta)W(\phi) \tag{RC4.2}$$

with the result that equation (RC4.1) becomes

$$\frac{VW}{r^2}\frac{\partial}{\partial r}\left(r^2\frac{\partial R}{\partial r}\right)+\frac{RW}{r^2\sin\theta}\frac{\partial}{\partial\theta}\left(\sin\theta\frac{\partial V}{\partial\theta}\right)+\frac{RV}{r^2\sin^2\theta}\frac{\partial^2 W}{\partial\phi^2}=0.$$

The first two terms depend on the variables r and θ but not ϕ; the third term depends on the variable ϕ but not on r and θ. It follows that we can separate the equation into the pair

$$\frac{\sin^2\theta}{R}\frac{\partial}{\partial r}\left(r^2\frac{\partial R}{\partial r}\right)+\frac{\sin\theta}{V}\frac{\partial}{\partial\theta}\left(\sin\theta\frac{\partial V}{\partial\theta}\right)=m^2 \tag{RC4.3a}$$

$$\frac{1}{W}\frac{\partial^2 W}{\partial\phi^2}=-m^2 \tag{RC4.3b}$$

where m is a constant to be assigned.

The second equation is that describing simple harmonic motion and we have

$$W(\phi)=\exp(\pm im\phi). \tag{RC4.4}$$

The solution must apply in the full range of angles from 0 to 180° and this means that m must take only integer values. Equation (RC4.3a) is readily separated into two equations with different variables by writing it first in the form

$$\frac{1}{R}\frac{\partial}{\partial r}\left(r^2\frac{\partial R}{\partial r}\right)=-\frac{1}{V\sin\theta}\frac{\partial}{\partial\theta}\left(\sin\theta\frac{\partial V}{\partial\theta}\right)+\frac{m^2}{\sin^2\theta}$$

and equating each side to the same constant, which it is convenient to write as $n(n+1)$.

This gives the two equations

$$\frac{1}{\sin\theta}\frac{\partial}{\partial\theta}\left(\sin\theta\frac{\partial V}{\partial\theta}\right)+\left[n(n+1)-\frac{m^2}{\sin^2\theta}\right]V=0 \qquad \text{(RC4.5a)}$$

and

$$\frac{\partial}{\partial r}\left(r^2\frac{\partial R}{\partial r}\right)+n(n+1)R=0. \qquad \text{(RC4.5b)}$$

Equation (RC4.5a) is called the Legendre equation. Its solution can be achieved in several ways but the expansion in powers of $\cos\theta$ is perhaps the most direct. The result is that $V(\theta)$ is written as the sum of terms of which the typical member is written

$$V(\theta)=P_n^m(\cos\theta)$$

where

$$P_n^m(\mu)=(1-\mu^2)^{m/2}(\mathrm{d}^m/\mathrm{d}\mu^m)P_n$$

and

$$P_n(\mu)=(1/2^n n!)(\mathrm{d}^n/\mathrm{d}\mu^n)(\mu^2-1)^n.$$

n and m are integers and $n\geqslant m\geqslant 0$. The functions P_n^m are called associated Legendre polynomials and the first members of the series are

$$P_1^1(\mu)=\sin\theta$$

$$P_2^1(\mu)=\tfrac{3}{2}\cos\theta\sin\theta$$

$$P_2^2(\mu)=\tfrac{3}{2}\sin^2\theta$$

$$P_3^1(\mu)=\tfrac{1}{2}\sin\theta(5\cos^2\theta-1)$$

$$P_3^2(\mu)=\tfrac{5}{2}\sin^2\theta\cos\theta$$

$$P_3^3(\mu)=\tfrac{5}{2}\cos^3\theta$$

$$P_4^1(\mu)=\tfrac{5}{8}\sin\theta\cos\theta(7\cos^2\theta-3)$$

$$P_4^2(\mu)=\tfrac{5}{8}\sin^2\theta(7\cos^2\theta-1)$$

$$P_4^3(\mu)=\tfrac{35}{8}\sin^3\theta\cos\theta$$

$$P_4^4(\mu)=\tfrac{35}{4}\sin^4\theta$$

for the first four sets of members. Further sets are usually not necessary in normal calculations. The functions P_n are called Legendre polynomials and the lowest-order members were listed earlier. Examples of the Legendre polynomials are shown in figure 4.3.

The solution of equation (RC4.5b) is supposed of the form r^a which is a solution of the equation if $a=-n(n+1)$ or n. Consequently,

$$r=a_1 r^n+a_2 r^{-(n+1)}$$

where a_1 and a_2 are two constants to be assigned. If the solution is to vanish at infinity we must set $a_1=0$; alternatively, if it is to vanish at the origin we set instead $a_2=0$.

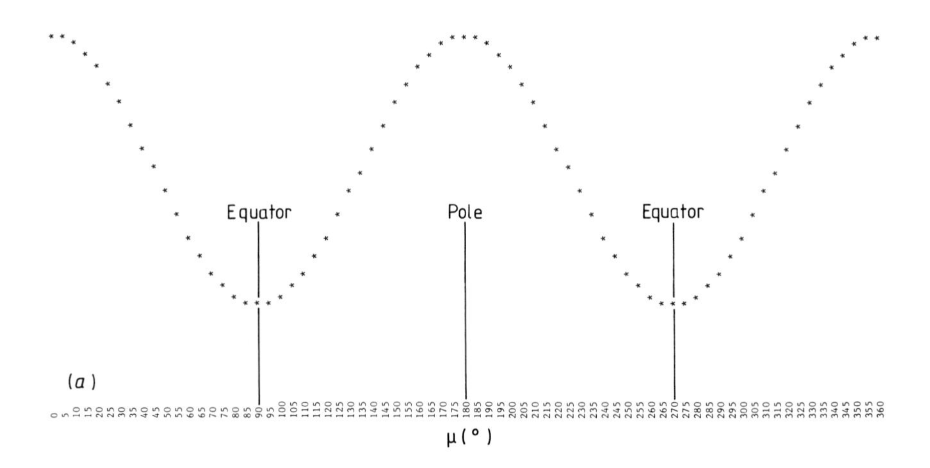

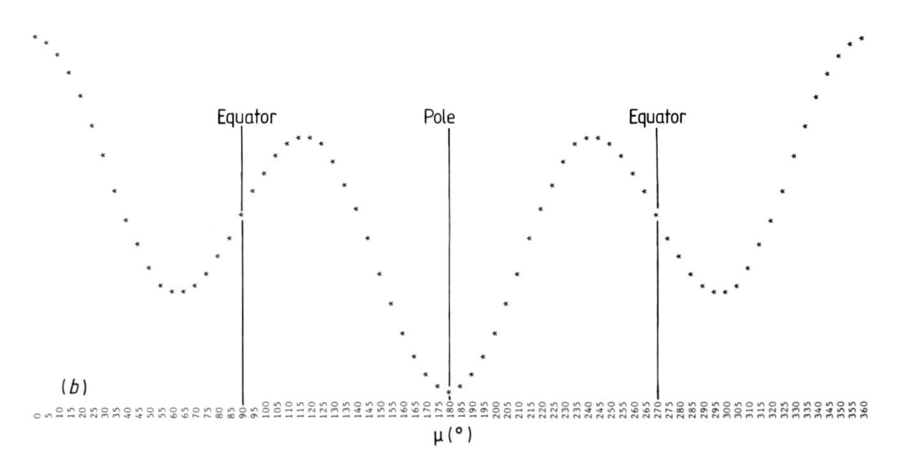

Figure 4.3 *Examples of Legendre polynomials:* (a) $P_2(\mu)$, (b) $P_3(\mu)$, (c) $P_4(\mu)$, (d) $P_2^1(\mu)$, (e) $P_3^1(\mu)$, (f) $P_4^1(\mu)$, (g) $P_2(\mu)P_3(\mu)\sin\theta$, *showing the zero integral for the normalisation condition* $\int_{-1}^{+1} P_2(\mu)P_3(\mu)\,d\mu$.

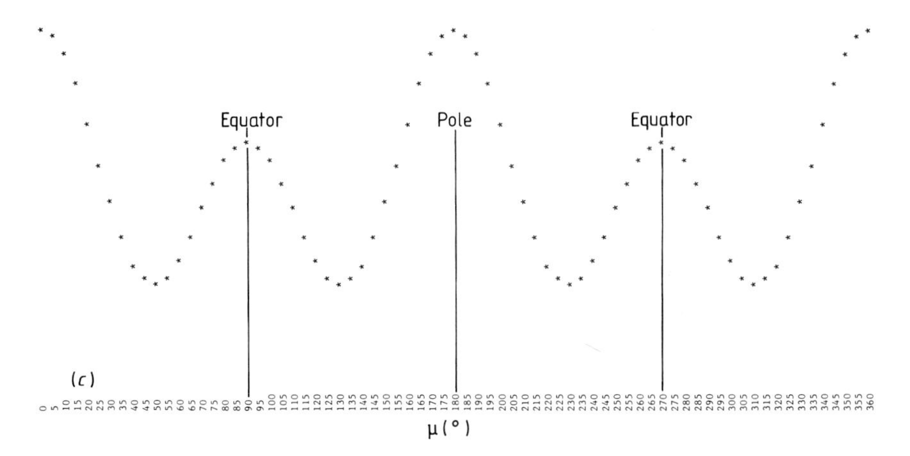
Equator
Pole
Equator
(c)
μ (°)

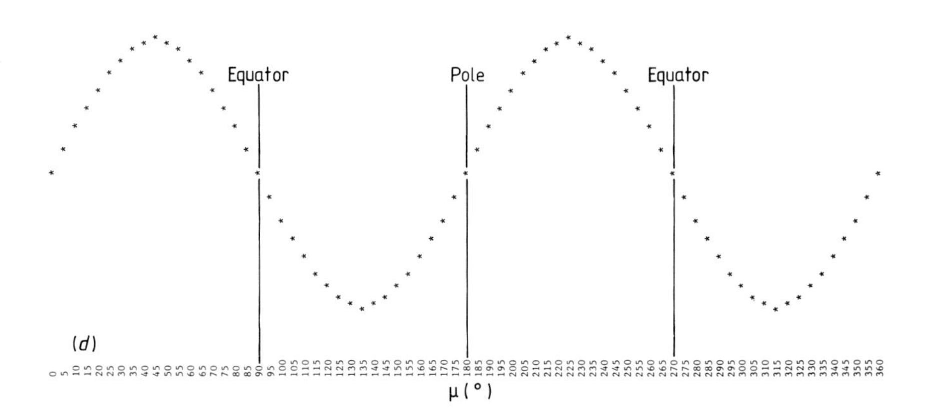
Equator
Pole
Equator
(d)
μ (°)

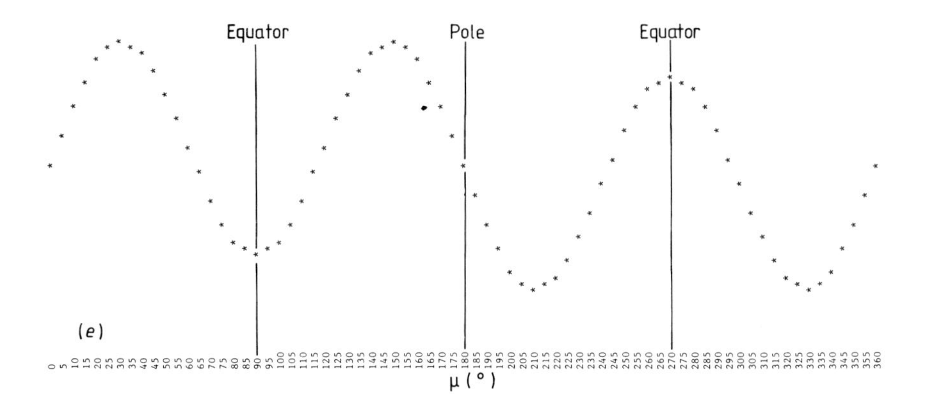
Equator
Pole
Equator
(e)
μ (°)

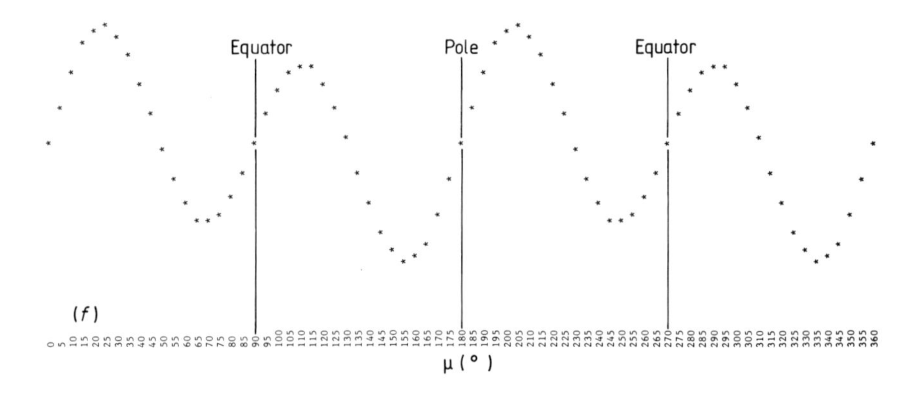

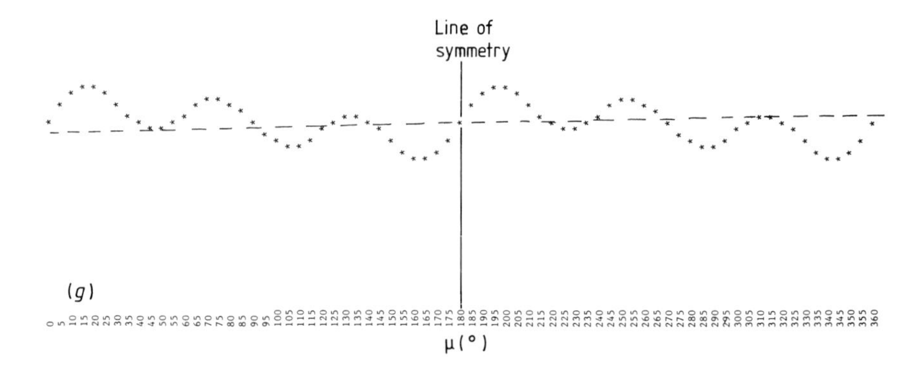

The separate components of the initial equation (RC4.1) are now known and the full solution can be set down in the form

$$U = \sum_{n,m}^{\infty} U_{nm} \tag{RC4.6a}$$

where

$$U_{nm} = (a_1 r^n + a_2 r^{-(n+1)}) P_n^m(\mu) \exp(\pm im\phi). \tag{RC4.6b}$$

The selection of the two constants is to be made according to the requirements of the boundary conditions. If longitude effects are not relevant the terms involving ϕ are eliminated and no associated functions P_n^m appear. If there is complete symmetry about the polar axes n can take even or odd values, but total reflection symmetry about the equatorial plane can only be represented by even values of n.

RC4.2 SHAPE OF THE EXTERNAL SURFACE

Just as an arbitrary period profile can be represented by the superposition of harmonic waves using the Fourier expansion method, so the arbitrary shape of a spheroidal surface can be represented by a particular superposition of Legendre polynomial curves. This is portrayed in figure 4.3 where individual polynomials are shown and the resultant profile which results from their superposition is also shown. It is seen that the representation of the surface in terms of the polynomials is a general and powerful way of showing complicated surface shapes.

There are two integral properties of the Legendre polynomials that are important to us in this connection. Remember that, since $\mu = \cos\theta$, μ is confined to values within the range $+1$ and -1. One relation is

$$\int_{-1}^{+1} P_n(\mu)P_m(\mu)\,\mathrm{d}\mu = (2/2n+1)\,\delta_{nm}$$

where δ_{nm} is the Kronecker delta symbol to be assigned the value $+1$ if $n=m$ but 0 otherwise.

Similarly,

$$\int_{-1}^{+1} P_n^m(\mu)P_1^m(\mu)\,\mathrm{d}\mu = [2(n+m)!/(2n+1)(n-m)!]\,\delta_{nl}$$

and such relations are often called normalisation conditions.

RC4.3 INERTIA FACTOR

The coefficient J_2 in the expansion (4.12) of the main text is related to the difference between the inertia factors in the polar and equatorial directions. This is shown as follows, and there are two stages to the argument.

RC4.3.1 Difference of two moments of inertia. The difference between the moment of inertia with respect to the polar axis and that with respect to an equatorial axis for a body is zero if the body is spherically symmetrical and not rotating, but is negative if the body is in rotation. This is readily appreciated when it is remembered that the rotating body has a smaller polar radius than equatorial radius, and consequently the material will be more concentrated in the polar direction than in the equatorial direction. The moment of inertia is smaller the greater the concentration of material towards the centre so that the polar moment will be smaller than the equatorial moment.

Consider a body with centre of mass O, polar (rotation) axis OX, and of total mass M_P and equatorial radius R_e. Let OE be a fixed direction in the equatorial plane, and so perpendicular to OX. This is displayed in figure 4.2. Consider a representative point P' within the body with polar coordinates (r', θ', ϕ'). The mass contained within a small element $\mathrm{d}V$ of volumes at P' is $\rho\,\mathrm{d}V$ where ρ is the local density.

The moment of inertia of the mass element $\rho\,\mathrm{d}V$ at P' about an axis of rotation is the product of the mass and the square of the perpendicular distance of the mass from the rotation axis. For the axis OX the perpendicular distance is $r' \sin \theta'$ whereas for the axis OE the perpendicular distance is $r' \sin \theta' \cos \theta'$. We assume the symmetry of the mass distribution to be such that ρ is not dependent on the longitude angle θ'. The moment of inertia I about the polar axis OX is

$$I = \int \mathrm{d}r' \int_0^{\pi} \mathrm{d}\theta' \int_0^{2\pi} \mathrm{d}\phi'(r')^2 \sin \theta' \rho(r')^2 \sin^2 \theta'$$

or

$$I = 2\pi \int \mathrm{d}r' \int_{-1}^{+1} \mathrm{d}(\cos \theta')\rho(r')^4 \sin^2 \theta'. \qquad \text{(RC4.7a)}$$

On the other hand, the moment of inertia A about the equatorial axis OE is

$$A = \int_V \mathrm{d}V\rho(r')^2(\cos^2 \theta' + \sin^2 \theta' \cos^2 \phi').$$

Carrying through the integration over the longitude angle ϕ' we find

$$A = 2\pi \int \mathrm{d}r' \int_{-1}^{+1} \mathrm{d}(\cos \theta')\rho(r')^4(\cos^2 \theta' + \tfrac{1}{2} \sin^2 \theta'). \qquad \text{(RC4.7b)}$$

The difference between I and A follows immediately from equations (RC4.7a) and (RC4.7b) as

$$I - A = -2\pi \int_0^{R_P} \mathrm{d}r' \int_{-1}^{+1} \mathrm{d}(\cos \theta')\rho(r')^4 P_2(\mu') \qquad \text{(RC4.8)}$$

where we have introduced the notation of the second Legendre polynomial.

RC4.3.2 Potential at a point outside the body. Consider now a point P with polar coordinates (r, θ, ϕ) outside the body so that $r > r'$ always. We wish to calculate the gravitational potential at P due to the body. The method is to find the potential due to the material at the representative point P' and sum all such contributions of all the representative points making up the body. We are concerned first with the potential distance PP' from P'.

If χ is the angle between the lines OP and OP' we have from trigonometry (see figure 4.2)

$$(PP')^2 = r^2 + r'^2 - 2rr' \cos \chi.$$

The potential $\mathrm{d}U_P$ at P due to the mass element $\rho\, \mathrm{d}V$ at P' is

$$\mathrm{d}U_P = -G\rho\, \mathrm{d}V(r^2 + r'^2 - 2rr' \cos \chi)^{-1/2}.$$

The total potential U_P at P is the sum of all $\mathrm{d}U_P$ throughout the volume, i.e.

$$U_P = -G/r \int_{V_P} \mathrm{d}V[1 + (r'/r) \cos \chi + \tfrac{1}{2}(r'/r)^2(3 \cos^2 \chi - 1) + \cdots].$$

The different terms of this expansion are evaluated as follows. The first term gives the total mass M_P according to

$$M_P = \int_{V_P} \rho \, dV.$$

The second term is evaluated by noting the result of spherical trigonometry

$$\cos \chi = \cos \theta \cos \theta' + \sin \theta \sin \theta' \cos(\phi - \phi')$$

from which we find, with ρ independent of ϕ' because of the symmetry of the mass

$$\int_{V_P} \rho r' \cos \chi \, dV = 0.$$

The third term is rearranged to the form

$$-2\pi G/4r^3 \int_{R_P} dr' \int_{-1}^{+1} d(\cos \theta')\rho r'^4 (3 \cos^2 \theta' - 1)(3 \cos^2 \theta - 1).$$

We now remember the result (RC4.8) to reduce this third contribution to a form involving the difference between the polar and equatorial moments of inertia:

$$U_P = -GM_P/R_P(R_P/r)[1 - (I - A)/\tfrac{1}{2}M_P R_P^2 (3 \cos^2 \theta - 1)] + O(R_P/r)^3.$$

Introducing J_2 by

$$J_2 = (I - A)/M_P R_P^2$$

and introducing the symbol P_2 we have the expansion for the potential at P

$$U_P = -GM_P/R_P(R_P/r)[1 - J_2(R_P/r)^2 P_2(\mu)] + O(R_P/r)^3.$$

This expression is called McCullagh's theorem.

RC4.4 FASTER ROTATION

The theory for a rapidly spinning planetary body has been developed by a number of authors. For an account and references see for example

Cook A H 1980 *Interiors of the Planets* (London: Cambridge University Press) (especially pp 249–59)

5

THERMAL EFFECTS AND THERMODYNAMICS

The overall equilibrium of a planetary body does not involve the thermal energy explicitly. Expressed another way, the mechanical equilibrium is compatible with a range of thermal states which can change without the overall equilibrium being lost. The thermal energy of the body is not zero, nor is it distributed uniformly throughout the volume initially. This implies a variation of temperature within the body, generally increasing with depth from the surface, and a surface temperature higher than that of the surroundings. Heat will flow within the body in such a way as to favour the condition of a uniform temperature throughout, and the planet will also lose heat by radiation to the surroundings. Some heat will be received from the Sun, though generally this makes an infinitesimal contribution to the total heat content. The thermal state of the planet will, therefore, change with time and the details of this thermal evolution are of the greatest interest as an indicator of the conditions inside. This approach has its most useful form as a comparative study of groups of planets both as a means of gaining an understanding of the physics of planetary bodies generally and as an indicator of the possible early history of the planetary system. There is here a very close link with the geology of the surfaces and physics, geology and chemistry overlap in these studies.

Measurements of the heat flow through the surface of the Earth have been made on a global scale over the last decade or so, and the description of the thermal Earth is beginning to achieve at least some limited success. The first steps have been taken to apply these arguments to other planets as well.

There are two primary sources of thermal energy of importance in planetary problems. One is the heat of formation of the body, presumably associated with some form of accretion process. The other arises from the decay of the small quantity of radioactive material in the body. To understand the thermal state of a body at a given time it is necessary to know details of both these contributions over the life of the body, but this is information that it is far from easy to find. The measurement of heat flow at the surface is not a sufficient basis to allow thermal conditions inside to be uniquely deduced because a wide range of combinations of accretional heat and radioactive heat can be constructed which lead to the same surface conditions. This is yet another example of the inverse problem in planetary physics where the direct problem of deducing the heat flow and temperature profile associated with a given distribution of heat sources, and which has a unique answer, is not the problem that faces us. Progress can be made on the basis of general principles of physics and it is these that will concern us in the present chapter.

5.1 RADIOACTIVE HEATING

Planetary material contains small quantities of radioactive impurity which decay according to the probability laws explained in RC1, equations (RC1.1) and (RC1.2). The radioactive emission can involve helium nuclei (alpha-particles), leptons (electrons, positrons and neutrinos of either parity: the electron–positron emission is called beta radiation), together with electromagnetic radiation (the nuclear energies placing the appropriate wavelengths in the gamma ray region). The electromagnetic radiation and neutrino emissions pass easily through the planetary material but the alpha and beta emissions do not. The surrounding planetary material stops the alpha- and beta-particles after a very short distance (the beta-particles after a few millimetres in rock and the alpha-particles more quickly still), the energy of motion of the emitted particles being shared with the atoms of the absorbing material. The emitted particle collides with atoms of the absorbing material and loses kinetic energy (and momentum) until it is brought to rest relative to the mean motion of the atoms of the absorber. In macroscopic terms the absorption of energy by the surroundings is seen as a small rise of its temperature. It is this transfer of radioactive energy to thermal energy of the surrounding material that provides the radioactive heating of the planetary interior.

There are four radioactive elements of importance for planetary heating. These are the two isotopes of uranium (${}_{92}U^{235}$, ${}_{92}U^{238}$), an isotope of thorium (${}_{90}Th^{232}$) and of potassium (${}_{19}K^{40}$). Other radioactive elements have proved useful as geochemical tracers (for example ${}_{37}Rb^{87}$) but these have little effect on the heat budget of the planet and so are not of interest here. The characteristics of the radioactive materials are contained in table 5.1.

It is seen that the radioactivity will provide a mean heating at the rate of 4×10^3 J kg^{-1} yr^{-1}. The concentration of the elements is small, typically about one part per million (ppm) of material for the Earth. The concentration for the

Table 5.1 *Heat resulting from radioactive decay of elements of geophysical interest.*

Isotope	Half-life (10^9 yr)	Heat ($J\ kg^{-1}\ yr^{-1}$)
${}_{92}U^{235}$	0.71	1.799×10^4
${}_{92}U^{238}$	4.50	2.970×10^3
U (natural)	—	3.054×10^3
${}_{90}Th^{232}$	13.90	8.366×10^2
${}_{19}K^{40}$	1.25	9.203×10^2
K (natural)	—	1.129×10^{-1}
${}_{37}Rb^{87}$	50.00	5.438×10^{-1}
Rb (natural)	—	1.505×10^{-1}

Moon is only little different and is likely to be of a similar magnitude for the other terrestrial planets. The heating of material containing this low level of concentration of radioactive materials is therefore generally about 4×10^{-10} J $kg^{-1}\ s^{-1}$ for naturally occurring rock.

This may be a very small amount of heat by ordinary standards but planetary conditions are different in two respects. First, the amount of material is large so the cumulative heating for the whole planet is correspondingly larger. Second, the half-lives of some of the materials involved are comparable with the age of the planet. Over the lifetime of the planet, radioactive heating will have provided some $4 \times 10^7\ J\ kg^{-1}$ of heat and this is not a small amount.

5.2 UNCERTAINTIES CONCERNING RADIOACTIVE HEATING

There are, at the present time, two major uncertainties associated with any assessment of the radioactive contribution to the heat budget of a planetary body.

One is the distribution and precise number of radioactive atoms. The surface conditions of the Earth, and the Moon at the Apollo sites, have been sampled but the amount in the regions below the surface can only be guessed at. It would seem that the cosmic abundance of elements would be seriously distorted if the surface radioactive content were typical of the whole volume and the interiors of the Earth and Moon would contain much more heat than seems possible on general grounds. Apparently, the surface region is much richer in radioactive material than the inside for reasons that are not yet understood. Although the distribution with depth is still to be determined, planetary models can be useful in assessing the limits of the distributions of radioactive materials that are consistent with independent considerations.

A second uncertainty is in the actual radioactive materials of relevance to the total heat balance of the planet. Materials with a half-life small compared with the age of the planet (which means compared with 4.5×10^9 years) will not be active now but could have provided substantial quantities of heat in the past, during the early life of the planet. This heat can still be locked up within the planet. One very strong candidate in this connection is ${}_{13}Al^{26}$ which has a half-life of about 7.2×10^5 years. The heating of the early Solar System could well have been affected in significant ways by the presence of this radioactive material.

5.3 COMMENTS ON THE GENERAL TRANSFER OF HEAT

A region initially hotter than its surroundings will lose heat to the environment by any of three mechanisms of heat transfer, although not all will usually be operative simultaneously (see RC5.1).

One mechanism is radiation; it is electromagnetic heat radiation which is propagated from a body at temperature T independently of whether there is matter there or not. For the ideal case of a black body, the flux of heat q (which is the heat flow per unit area through a (possibly) hypothetical surface) is proportional to the fourth power of the absolute temperature T at the surface. This is Stefan's law of radiation. Explicitly:

$$q = aT^4 \tag{5.1}$$

where a is the universal constant of Stefan ($a = 7.55 \times 10^{-16}$ J m^{-2} K^{-4}). We are familiar with heat and light radiation from the Sun, the surface of which is very closely a black body at 5800 K.

The second mechanism is that of the simple conduction of heat. This requires material to transmit the energy and, for a rigid conductor of heat, the heat flux q is proportional to the space gradient of the temperature along the direction of the flow. Then

$$q = -k\,\partial T/\partial n \tag{5.2}$$

in the direction n of the flow. Here k is the coefficient of heat conduction and the negative sign follows because heat flows from the hotter to the colder region. For a homogeneous material, k is a characteristic of the material itself and is independent of the gradient of temperature. This is Fourier's law. Equation (5.2) represents the observed situation very well for sufficiently small temperature gradients and for small variations of pressure and ambient temperature. k is actually weakly dependent on temperature and pressure and this can be significant in planetary problems where extended volumes of the interior of a larger planetary body are involved.

The third mechanism of heat transfer is found in a fluid heat conductor in a gravitational field and which also carries a temperature difference across it. If the flow of heat is greater than a characteristic magnitude, the carriage of heat arises from the movement of the fluid itself. The material moves in an ordered way to carry the heat according to the laws of conservation of energy and fluid momentum. This is the natural convection of the fluid.

Each of the three mechanisms has a part to play in the thermal evolution of a planetary body. The relative importance of each mechanism and the effects it has on the internal conditions will depend upon the particular size and composition of the planetary body.

5.4 PASSAGE OF HEAT BY CONDUCTION

The description of the conduction of heat in a solid body is based on the well established conservation of energy and considers the balance of heat in a chosen unit volume of material.

The heat content of the volume will change with time as the result of two separate processes: one is the passage of heat through the surface enclosing the volume (either in or out); the other is the production of heat inside. These are expressed mathematically by the expression

$$\rho c_p \, \partial T/\partial t = \operatorname{div} q + H_v \tag{5.3}$$

where ρ is the density, c_p the specific heat at constant pressure, q the heat flux and H_v the heat produced per unit of volume per unit time. For our present purposes H_v is usually due to radioactive decay.

In applying this equation in practice it is necessary to express q in terms of the temperature to provide a mathematical problem with one unknown to be found. For a simple material, we apply the Fourier law statement (5.2) in its general vector form

$$q = -k \operatorname{grad} T.$$

The sign convention must be accounted for. Heat is *lost* to the volume by conduction and a positive sign in equation (5.3) means heat gained; div q must, therefore, carry a negative sign. For the loss of heat to the volume, therefore,

$$-\operatorname{div} q = -\operatorname{div}(-k \operatorname{grad} T) = \operatorname{div}(k \operatorname{grad} T).$$

For a homogeneous heat conductor, k is constant and

$$\begin{aligned} \operatorname{div}(k \operatorname{grad} T) &= k \operatorname{div} \operatorname{grad} T \\ &= k \nabla^2 T \end{aligned} \tag{5.4}$$

where ∇^2 is the Laplacian. For cartesian coordinates (x, y, z)

$$\nabla^2 = \frac{\partial^2}{\partial x^2} + \frac{\partial^2}{\partial y^2} + \frac{\partial^2}{\partial z^2} \tag{5.5a}$$

while for a spherical symmetrical geometry (radius r, co-latitude θ and longitude ϕ)

$$\nabla^2 = \frac{1}{r^2}\frac{\partial}{\partial r}\left(r^2 \frac{\partial}{\partial r}\right) + \frac{1}{r^2 \sin\theta}\frac{\partial}{\partial \theta}\left(\sin\theta \frac{\partial}{\partial \theta}\right) + \frac{1}{r^2 \sin^2\theta}\frac{\partial^2}{\partial \phi^2}. \quad (5.5b)$$

Inserting equation (5.4) into equation (5.3) gives

$$\rho c_P \, \partial T/\partial t = k \, \nabla^2 T + H_v \quad (5.6)$$

as the equation for the distribution of temperature for a homogeneous conductor. Notice that the distribution of temperature in space at a given time is linked to the time evolution of the distribution.

It is convenient to divide equation (5.6) by ρc_p, introducing the thermometric conductivity $K = k/\rho c_p$ and $h = H_v/\rho c_p$. The result is

$$\partial T/\partial t = K \, \nabla^2 T + h. \quad (5.7)$$

We can notice that the spatial dependence of T enters the equation through the second derivative while the dependence on the time enters through the first derivative. This type of mathematical equation describes irreversible processes, in this case the movement of the system from an initial non-uniform distribution of heat energy to the final state of a uniform distribution. Equation (5.7) has the form of a diffusion equation.

It is convenient to introduce the dimensionless variables ξ, τ and θ according to

$$\xi = \frac{r}{R_P} \qquad \tau = \frac{Kt}{R_P^2} \qquad \theta = \frac{T}{T_m} \quad (5.8)$$

where R_P is the total planetary radius and T_m the melting point of the material. Spherical symmetry is often appropriate to planetary problems, and for this case equation (5.7) takes the dimensionless form

$$\frac{\partial \theta}{\partial \tau} = \frac{1}{\xi^2}\frac{\partial}{\partial \xi}\left(\xi^2 \frac{\partial \theta}{\partial \xi}\right). \quad (5.9)$$

For a solid, $T < T_m$; indeed we can write $T < \gamma T_m$ where γ is a numerical factor probably less than about 0.7 to avoid the plastic state near the melting point.

5.5 TIME SCALE FOR CONDUCTION

Consider a sphere with no internal heating. The time for the heat to conduct a distance L is, according to equation (5.8), $t \sim (L^2/K)$. For terrestrial-type material, $K \sim 10^{-3}$ m^2 s^{-1}. Then for $L = 1$ km, $t = 10^{12}$ s $= 10^3$ yr, while for $L = 1000$ km, $t = 10^{18}$ s $= 10^{11}$ yr. This is an astonishingly long time compared with the age of the Earth ($= 4.5 \times 10^9$ yr) and shows that the passage of heat by conduction is a very slow process. For $t=4.5\times10^9$ yr, $L=100$ km or not much deeper than the terrestrial crust. For planetary bodies larger than this the initial heat will not yet have had time to conduct to the surface, the primordial heat still being locked inside the planet. It is interesting that this depth is about the magnitude of the minimum radius for a planetary body according to the arguments of chapter 2.

5.6 SURFACE BOUNDARY CONDITIONS

The solutions of the temperature equation relevant to a particular physical system are selected by means of boundary conditions. There are several aspects.

It is supposed that the observed temperature distribution at the time instant t is related directly to the initial distribution. Any solution of equation (5.7) for the time t must reduce to the given initial distribution as $t\rightarrow0$. If the initial distribution is discontinuous this condition is to apply to the continuous regions of the temperature field.

The surface conditions are important and various forms can be of interest. Thus, the surface temperature may be constant, or be a function of the time or location or both together. It may be that there is no heat flux across the surface so that

$$\frac{\partial\theta}{\partial n}=0 \qquad \text{everywhere on the surface.} \tag{5.11a}$$

For a prescribed flux across the surface the condition is easily amended. This will lead to a loss of heat to the outside described by a radiation boundary condition. If the surface were a black body the loss of heat could be represented by the condition

$$k\frac{\partial\theta}{\partial n}=a(T_s^4-T_o^4) \tag{5.11b}$$

where T_s is the surface temperature and T_o the temperature of the environment. For many cases of interest $T_s\gg T_o$ and $(T_s^4-T_o^4)$ can be well approximated by T_s^4. For a more general surface we replace the radiation constant by $a\varepsilon$ where ε is the emissivity of the surface. For a polished metal surface $\varepsilon\sim2\times10^{-2}$ while for soil it may rise to 8×10^{-1}.

If the surface is enclosed by an atmosphere in motion a linear condition is sometimes relevant. The boundary condition can now be

$$k\frac{\partial\theta}{\partial n}=\mathscr{H}(T_s-T_o) \tag{5.11c}$$

where $\mathscr{H}$ is a surface heat transfer coefficient. It is not strictly a constant but depends on the precise surface conditions and, to an extent, on the temperature.

The heat flux through the surface separating two different media is continuous. If θ_1 and θ_2 are the two non-dimensional temperatures and K_1 and K_2 the corresponding conductivities then

$$-K_1 \frac{\partial \theta_1}{\partial n} = K_2 \frac{\partial \theta_2}{\partial n} \tag{5.11d}$$

where n is along the direction normal to the two surfaces.

The magnitude of the loss of heat at the surface to the surroundings is difficult to assess over an extended time period but fortunately it makes only a small contribution to the internal heat balance. It is, of course, of central importance in determining the precise equilibrium conditions in a crust of given composition and its specification is vital to an understanding of the early history and evolution of the crustal material. Of particular interest here is the process of solidification of material and this aspect could act as a guide in particular cases between cold or hot initial accretion. This matter is considered in RC5. The general result is found that silicate-type material forming a crust and initially molten at the melting point will require as little time as 10^8 years to solidify to a depth of about 100 km in the absence of short lifetime radioactive materials. It may be more than coincidence that depths of this order are found to be special in the internal structures of the Earth and the Moon.

5.7 ADIABATIC TEMPERATURE GRADIENT

The temperature within a planet increases with depth due to compression and the temperature gradient can be found from thermodynamic arguments. This temperature gradient must act as a standard against which any thermal effects of radioactivity are to be judged.

Consider a unit mass of material and suppose a quantity $\mathrm{d}Q$ of heat energy enters the volume. The internal energy U increases by an amount $\mathrm{d}U$ and the volume V changes by $\mathrm{d}V$ such that

$$\mathrm{d}Q = \mathrm{d}U + p\,\mathrm{d}V \tag{5.12}$$

with p being the pressure. This is written, alternatively,

$$\mathrm{d}Q = \mathrm{d}(U + pV) - V\,\mathrm{d}p \tag{5.13}$$

and $H = U + pV$ is the enthalpy or total heat.

The enthalpy is a function of the pressure and temperature so

$$\mathrm{d}H = \left(\frac{\partial H}{\partial T}\right)_p \mathrm{d}T + \left(\frac{\partial H}{\partial p}\right)_T \mathrm{d}p.$$

But $(\partial H/\partial T)_p = c_p$, the heat capacity at constant pressure, so equation (5.13) is alternatively

$$dQ = c_p\, dT + \left[\left(\frac{\partial H}{\partial p}\right)_T - V\right] dp.$$

It is a result of thermodynamics that

$$\left(\frac{\partial H}{\partial p}\right)_T - V = -T\left(\frac{\partial V}{\partial T}\right)_p$$

so that

$$dQ = c_p\, dT - T\left(\frac{\partial V}{\partial T}\right)_p dp. \qquad (5.14)$$

The magnitude of dQ depends on the physical process of heat transfer. Of particular interest is the case where there is no heat flow ($dQ = 0$). This condition is one of adiabatic compression. Then equation (5.14) becomes

$$\frac{dT}{dp} = -\frac{T}{c_p}\frac{\partial V}{\partial T}.$$

The coefficient of the volume expansion β is defined by

$$\beta = -\frac{1}{V}\frac{\partial V}{\partial T}$$

and (because we are dealing with unit mass) $V = 1/\rho$. Then

$$\frac{dT}{dp} = \frac{T}{c_p}\frac{\beta}{\rho} \qquad (5.15)$$

so that, if z is the depth

$$\frac{dT}{dp} = \frac{dT}{dz}\frac{dz}{dp} = \frac{T}{c_p}\frac{\beta}{\rho}.$$

From the condition of hydrostatic equilibrium

$$dp/dz = \rho g$$

and equation (5.15) is finally

$$\frac{dT}{dz} = \frac{g\beta T}{c_p}. \qquad (5.16)$$

The gradient (dT/dz) is the adiabatic temperature gradient.

Its importance lies in its neutral thermal behaviour. To explain this, consider two small elements of mass dm_1 and dm_2 in a compressible fluid in a permeating gravitational field, with dm_1 vertically above dm_2. The temperature and pressure both increase with depth so the magnitudes of these variables associated with dm_2 will both be higher than those associated with dm_1. If we perform a hypothetical experiment in which dm_1 and dm_2 were interchanged instantaneously, dm_2 would have a higher temperature and pressure than its environment while the magnitudes for dm_1 would be too low. The mass dm_2 would expand with a fall of temperature while dm_1 would contract with a rise of temperature. If the temperature gradient within the bulk fluid is adiabatic the temperatures and pressures of dm_1 and dm_2 would simply be interchanged. If the gradient is larger than the adiabatic it is said to be superadiabatic, while if it is smaller it is subadiabatic. For a superadiabatic condition such as we have described the density will be higher than the local value and the local fluid element will rise. Otherwise it will sink.

The adiabatic gradient provides a criterion for mechanical stability of the fluid and a superadiabatic gradient will be associated with a motion which is called natural convection.

5.8 NATURAL FLUID MOTIONS

The material of most of the interior of a planetary body is unable to withstand transverse stresses over long periods of time and in this sense can be regarded as a liquid with a very high shear viscosity. The physics of such material is not known with any precision but it is believed that the usual principles of fluid mechanics will apply in a general sense with the viscosity taken as very large. It is of interest, on this basis, to notice the essential features of the physics of fluids for application to planetary problems.

A fluid moves under the action of forces in a way consistent with the conservation of mass, momentum and energy in the fluid. The equation (5.7) is a particular form of the energy equation for a solid heat conductor and can be readily extended to include fluid flow. Each fluid element will carry heat $\rho c_p \boldsymbol{v}$ as it moves with velocity $\boldsymbol{v}$ and, combined with the convective flow, gives a total heat flux

$$\boldsymbol{q} = \rho c_p \boldsymbol{v} - k \text{ grad } T. \tag{5.17}$$

The heat balance equation (5.7) becomes now

$$\frac{dT}{dt} = K \nabla^2 T - h \tag{5.18a}$$

where

$$\frac{d}{dt} = \frac{\partial}{\partial t} + (\boldsymbol{v} \cdot \text{grad}) \tag{5.18b}$$

is the total derivative moving with the fluid. Equations (5.18) are an expression of the resultant effect of conducted and convected heat and heat sources.

Orders of magnitude are interesting. For a steady temperature distribution without heat sources and for fluid flow in the x-direction with characteristic speed U, heat will diffuse typically in the z-direction and equation (5.18a) gives the relation

$$U/x \sim K/z^2.$$

The distance diffused by the heat for a given distance of flow downstream is

$$z \sim \sqrt{\frac{Kx}{U}} = \sqrt{\frac{K}{\nu}}\sqrt{\frac{\nu}{Ux}}\,x \tag{5.19}$$

where we have introduced the effects of the shear viscosity through the kinematic viscosity $\nu = \mu/\rho$, μ being the coefficient of shear viscosity and ρ the fluid density.

The expression for the conservation of mass follows from arguments analogous to those introduced in §5.4 for the heat conduction equation but with the difference that there are no sources of mass analogous to sources of heat. The mass flux is $\rho\boldsymbol{v}$ and the expression for continuity is

$$\frac{\mathrm{d}\rho}{\mathrm{d}t} + \rho \operatorname{div} \boldsymbol{v} = 0$$

for unit volume, called the continuity equation.

The conservation of momentum is derived by applying Newton's laws of motion to a small fluid volume. Then we have

$$\rho\frac{\mathrm{d}\boldsymbol{v}}{\mathrm{d}t} = \boldsymbol{F} \tag{5.20}$$

where $\boldsymbol{F}$ is the sum of the forces acting on the volume and the left-hand side is the inertia associated with the moving fluid element; $\boldsymbol{F}_\mathrm{I} = \rho(\mathrm{d}\boldsymbol{v}/\mathrm{d}t)$ is the inertia force. There are several contributions to $\boldsymbol{F}$ to take into account: the pressure force $\boldsymbol{F}_\mathrm{p}$, the force of gravity $\boldsymbol{F}_\mathrm{g}$, the rotation force $\boldsymbol{F}_\mathrm{r}$ and the viscous force $\boldsymbol{F}_\mathrm{v}$. Then

$$\boldsymbol{F} = \boldsymbol{F}_\mathrm{p} + \boldsymbol{F}_\mathrm{g} + \boldsymbol{F}_\mathrm{r} + \boldsymbol{F}_\mathrm{v}. \tag{5.21}$$

The effects of these various forces are compared conveniently by making order of magnitude estimates for them. Thus

$$F_\mathrm{I} = |\rho\, \mathrm{d}\boldsymbol{v}/\mathrm{d}t| \sim \rho U^2/L \tag{5.22}$$

for a fluid region of characteristic dimension L. The pressure force due to the flow is

$$F_\mathrm{p} = -|\mathrm{grad}\; p| \sim p/L$$

while the gravity force is

$$F_\mathrm{g} = |\rho\boldsymbol{g}|. \tag{5.23}$$

The force due to the rotation of a fluid with angular velocity w with respect to a stationary observer is

$$F_r = |\rho[\boldsymbol{v} \times \boldsymbol{w}]| \sim Uw\rho \tag{5.24}$$

and finally the viscous force for an incompressible flow (almost always a sufficient approximation in planetary interiors) is

$$F_v = |\mu \nabla^2 \boldsymbol{v}| \sim \mu U^2/L^2 \tag{5.25}$$

where μ is the coefficient of shear viscosity.

Equations (5.18) and (5.20) are coupled via the density and flow velocity. It is usually sufficient to account for the effect of temperature on the density in equation (5.20) only through the term involving gravity, the density otherwise being assigned a mean magnitude independent of temperature (Boussinesq approximation). Writing β for the volume coefficient of temperature and ρ for the density then for a temperature difference $\Delta T = T - T_o$ which is not too large

$$\rho = \rho_o(1 - \beta\, \Delta T).$$

ρ_o is the density corresponding to the temperature T_o. Then equation (5.23) becomes alternatively

$$\begin{aligned} F_g &= g\rho = g\rho_o(1 - \beta \Delta T) \\ &= g\rho_o - g\rho_o \beta\, \Delta T. \end{aligned}$$

The first term is independent of the temperature: the second term represents the buoyancy force F_B so that

$$F_B = |g\rho_o\beta\, \Delta T|.$$

The density in each term of the equation of motion is now ρ_o but it is convenient not to write the subscript o on the understanding that a mean density is involved.

The relative effectiveness of the various forces in determining the flow characteristics is assessed by comparing the strengths of the forces. This is conveniently done through order of magnitude estimates and the ratio of any two expressions is a dimensionless quantity, i.e. a pure number. A range of 'numbers' (well known in fluid mechanics) follow at once. With the strength of any one force taken as standard (often the inertia force but not always) the relative strengths of the others are expressed in numerical terms, as is explained in RC5. Conditions at the boundary are also included, and reference should be made to these at this point.

5.9 CONDITIONS PECULIAR TO GEOPHYSICS

The application of these arguments to particular planetary interiors is controlled by the very large magnitude of the kinematic viscosity suggested by observation. This is typically $\nu \sim 10^{16}\ \mathrm{m^2\,s^{-1}}$. With $K \sim 10^{-6}\ \mathrm{m^2\,s^{-1}}$ we obtain a Prandtl number $Pr \sim 10^{22}$ and Reynolds number $Re \sim 10^{-19}$, based on a continental length scale of $L = 10^3$ km. Pr is larger and Re smaller than the corresponding laboratory magnitudes by many orders of magnitude. This can usually be accounted for in calculations by supposing Pr to be indefinitely large and the inertia force is not a controlling factor for the flow.

Natural motions are caused by differences of density at different depths in the presence of gravity, and for thermal conditions these differences are caused by temperature variations. For vertical motion this difference must be in excess of the local adiabatic temperature increment due to compression alone. The buoyancy force is now operative, opposed by the viscous force. Motion is caused by the unbalance of these two forces. For motion, the buoyancy force must dominate so that

$$g\rho\ \Delta T > \mu U/L^2 > g\rho(\mathrm{d}T/\mathrm{d}L)_{\mathrm{A}}L$$

where $(\mathrm{d}T/\mathrm{d}L)_{\mathrm{A}}$ is the appropriate adiabatic gradient.

In terms of the Rayleigh number Ra (see equation (RC5.14g), p 130), motion requires that

$$Ra > RePr. \tag{5.26}$$

Pr is very large and Re is very small with a magnitude determined by that of the distance scale L, so equation (5.26) is not a precise criterion. We can infer that Ra must exceed some critical magnitude Ra_{c} and laboratory experiments (of course using simple fluids) suggests that $Ra_{\mathrm{c}} \sim 10^3$. It is interesting that this result is consistent with planetary material of scale $L \sim 10^3$ km.

The initial planetary heat of formation supplemented by subsequent radioactive heat can escape to the surface only slowly by conduction. The consequence of such an accumulation of heat is a rise in the temperature of the material. The rigidity of the solid material will fall (or for creep the viscosity will decrease) and motion will be possible over geologically short time periods (of the order of 10^6 yr with the material behaving as a fluid of high viscosity. The motion will transport heat with the consequence that the temperature locally will fall. The viscosity is very sensitive to temperature, with a dependence of the form $\exp(B/T)$, where B is an activation energy and T the local temperature, and a fall in temperature produces a marked increase in the viscous resistance to flow. Slower flow means less heat transport and so a rise of temperature, which causes the viscosity to decrease, and so on. An equilibrium can be anticipated in which the viscous dissipation assumes a magnitude controlled by the thermal conditions.

These are qualitative arguments based on the known behaviour of Newtonian liquids, but planetary material will not behave like such a simple liquid. While this is undeniably the case, and while detailed laboratory experience is largely irrelevant for planetary studies, the general relationships deduced this way are generally regarded as having a wider relevance for complicated liquids like creeping rock material. In particular, the development of convective regimes in the planetary material can be expected and the conditions for the onset of particular fluid flow patterns are taken to be independent, to a large extent, of the precise nature of the liquid. This means that the importance of the Rayleigh number is still valid and the critical magnitude is accepted as being a useful guide to actual conditions.

5.10 NATURAL CONVECTION

The motion of the liquid once buoyancy effects become important is a circular motion in the vertical plane (involving vorticity with a single component in the horizontal direction for the simplest cases) in which hot fluid rises and cold fluid falls. In laboratory experiments the vertical extent L_v of the motion is comparable in magnitude to the horizontal scale L_h. The ratio $L_v/L_h = \alpha$ is called the aspect ratio and is equal to about one for laboratory conditions. The aspect ratio for planetary materials is unknown, but if it is unity, planetary convection will involve a depth scale comparable to the surface extent.

The vertical motion is driven by buoyancy forces and opposed by friction. The horizontal motion linking the hot upward flow to the cold downward flow is determined by conditions of mass continuity. The vertical motion transports heat upwards and is isothermal, to a good first approximation. The horizontal motion at the top and bottom has the nature of a thermal boundary layer. Heat is passed from the fluid to the colder material above at the top while heat passes from the hotter material below into the fluid at the bottom. These ideas can be expressed in terms of the forces acting on the fluid.

For the horizontal motion, top and bottom, the balance of forces comes from the equation for the temperature; for steady conditions

$$v_h \frac{\partial T}{\partial x} \sim K \frac{\partial^2 T}{\partial z^2}$$

where z is the vertical direction and x the horizontal. In general terms, $x \sim L_h$ so that

$$\frac{U_h}{L_h} \sim \frac{K}{\delta^2} \qquad (5.27a)$$

where δ is the thickness of the thermal boundary layer.

For the vertical motion the balance of momentum gives

$$\beta g \rho \, \Delta T \sim \mu \frac{\partial}{\partial z} \frac{\partial v_x}{\partial x}$$

or, to an order of magnitude (with $z \sim L_v$)

$$\beta g \rho \, \Delta T \sim \mu (U_v/L_v)\delta \qquad (5.27b)$$

where ΔT is the temperature difference across the fluid layer.

It follows by rearrangement that

$$\delta \sim (Ra)^{-1/3} L_{\mathrm{h}}$$

$$U_{\mathrm{v}} \sim U_{\mathrm{h}} \sim (Ra)^{2/3} K/L_{\mathrm{h}}$$

$$q_{\mathrm{conv}} \sim \rho c_{\mathrm{p}} U_{\mathrm{v}} \left(\frac{\Delta T}{L_{\mathrm{h}}}\right) \delta \sim (Ra)^{1/3} \qquad (5.28)$$

where q_{conv} is the convective heat flux. It is natural to involve the Nusselt number Nu (equation (RC5.14l), p 132) in the specification of the heat flow at the surface and we find

$$Nu \sim (Ra/Ra_{\mathrm{c}})^{1/3}$$

where Ra_{c} is the critical Rayleigh number. If we can take $Ra_{\mathrm{c}} \sim 10^3$ and $Ra \sim 10^6$ (probably not untypical of terrestrial conditions), $Nu \sim 10$, showing the heat flow by convection to be greater than that by conduction by the factor 10. Referring to the arguments of §5.7 we see that even convective motion of this magnitude will not be sufficient to allow heat to have dissipated over the present life span of a planet of terrestrial size.

The time t_{c} for material to make one circuit of convection is deduced from equation (5.27b). We have

$$U_{\mathrm{v}} \sim L_{\mathrm{v}}/t$$

so that

$$t \sim \frac{\nu}{\beta g \, \Delta T L_{\mathrm{v}}}$$

$$= \left(\frac{\nu K}{\beta g \, \Delta T L_{\mathrm{v}}^3}\right) \frac{L_{\mathrm{v}}^2}{K}$$

and hence

$$t \sim \left(\frac{1}{Ra}\right) \frac{L_{\mathrm{v}}^2}{K}.$$

This will be the time for each quarter of the cycle (one up, one down and two horizontal) so that $t_{\mathrm{c}} \sim 4t$. For terrestrial mantle conditions and for $L_{\mathrm{v}} \sim 10^3$ km we find $t \sim 5 \times 10^7$ yr so that $t_{\mathrm{c}} \sim 2 \times 10^8$ yr. This time decreases as Ra increases, i.e. as the intensity of convection increases. For an object of lunar size ($g \sim 2$ m s^{-2}) this estimate is *increased* by a factor 5, the convective motion being less vigorous than for the Earth because the intensity of gravity is lower. If the combination of gravitational intensity and temperature difference is large enough, the ordered convective motion will dissolve into a disordered (turbulent) form. The expressions (5.28) will then be replaced by others and, from experience in other areas of fluid mechanics, the dependence of Nu on Ra might be expected to have the form $Nu \sim (Ra/Ra_{\mathrm{c}})^{1/4}$. This is a weaker dependence than for ordered motion but will carry more heat because the conditions are further from equilibrium.

5.11 CONDITIONS FOR CONVECTION

It is relevant to ask whether convective motion will actually arise in planetary material having every appearance of being solid on a day-to-day basis. The answer hinges on whether the viscosity is such as to allow motion over time scales shorter than those controlled by conductive heat transfer processes.

Approach the question by way of the numbers considered earlier. We are concerned with the ratio (which is a number) of the convective heat to the conductive heat transports within the body of the planet. For conduction alone, this number, called the Peclet number Pe (equation (RC5.14j), p 131), is unity; otherwise $Pe > 1$. We write

$$Pe = |\boldsymbol{v} \cdot \text{grad } T| / |K \nabla^2 T| \sim K/UL$$

and we see at once that $Pe = PrRe$. For the thermal conditions we have been considering $Pr \sim 10^{22}$ and $Re \sim 10^{-19}$ so that $Pe \sim 10^3$, and this certainly satisfies the condition $Pe > 1$. We conclude that the conditions are right for convective flow, in principle at least.

The conditions of fluidity of a material can be assessed by the response time to the action of forces. Within a planetary body the equilibrium results from the action of pressure and viscous forces. We have seen in a previous chapter that the transverse forces, i.e. the transverse stresses, in the material are negligibly small within the major part of the volume, a characteristic of fluids. The balance between pressure and viscous forces is expressed by

$$\text{grad } p \sim \mu \nabla^2 \boldsymbol{v}$$

or to an order of magnitude

$$p/L \sim \mu U/L^2. \tag{5.29}$$

This gives for the speed

$$U \sim pL/\mu \sim L/\tau$$

where τ is the time for the flow. Then

$$\tau \sim \mu/p.$$

For the Earth, at a depth of 10^3 km, $p \sim 3 \times 10^{10}$ N m^{-2} and $\mu \sim 10^{20}$ Pa. This gives $\tau \sim 10^{10}$ s $\sim 10^3$ yr. For a depth of 10 km, this time is raised to 10^5 yr, consistent for instance with the deduced time scale for movement below the Scandinavian shield, due to the release of loading as ice laid down during the last ice age has melted. The movement of terrestrial material due to pressure forces is surprisingly fast, which is another way of saying that the material moves as a fluid. The fluidity is greater the smaller the viscosity, i.e. the higher the temperature. Neglecting radiation at the surface (reasonable because of the slowness with which heat is transported to the surface—material below about 10^2 km is virtually uncoupled from the surface anyway) the rise in temperature due to the radioactive heat produced over the time t_r is given by

$$ht_r \sim \rho c_p \, \Delta T.$$

If h and t_r are large enough, the material can be raised to its melting temperature. This is the more possible when initial heat of compression is also taken into account. Conditions are then right for convection to occur and there can be little doubt that convection will arise in every planetary body sooner or later, and sooner rather than later. Such motion will persist until the heat content has virtually disappeared, and for a planet of terrestrial type this time will be of the order 10^{10}–10^{11} yr.

We have said nothing so far about rotation. The ratio of the inertia to rotation forces is measured by the Rossby number (see equation (RC5.14h), p 130) and for rotation to be negligible this number must be large in comparison with unity. The magnitude in a particular case will depend on the nature of the planetary body.

For a terrestrial-type body with very large viscosity the material speed is extremely small ($U \sim 10^{-9}$ m s^{-1}). Even if L is as low as 10^2 km the ratio $U/L \sim 10^{-14}$ and rotation speeds measured in days or more will provide $Ro > 1$ by many orders of magnitude. In these cases the effects of rotation are entirely negligible. These are the cases particularly relevant to the arguments developed in this chapter.

The situation is very different for giant fluid planets in rapid rotation (explicitly Jupiter and Saturn, with mean rotation periods of some 10 h), for now the viscosity of the constituent material is low. Then Ro can be of order unity and rotation effects need not be negligible within the main body of the planet. We shall have more to say about this in chapter 7.

5.12 COMMENT ON SURFACE CONDITIONS

The region immediately near the surface (say within 50 km of it) will make no essential contribution to the equilibrium inside but is important because it involves the part we actually see and perhaps walk on. For a silicate material radiation will quickly reduce the temperature below the melting point and conditions can be quite complicated.

The surface can be plastic (this will involve an appropriate temperature because the pressure is relatively low) and convective motion a possibility. The aspect ratio will need to be small ($\alpha \sim 10^{-2}$). Less dense material will rise and form a thin layer (continental crust) floating on more dense material (oceanic crust). The cooling of the continental material may be sufficiently fast for the density to rise well above the local value and so sink (subduction). Then a plate tectonic system can arise. This is not a general effect but arises only under special conditions. The surface conditions of two planetary bodies that might seem similar (for example Earth and Venus) may be different in detail but the conditions on each of the planetary bodies themselves may differ from one epoch to another (for instance plate tectonics is operative now on the Earth but this may not always have been so).

5.13 VARIATION OF PARAMETERS WITH DEPTH

As the depth increases, so the pressure and temperature will rise, and it may be necessary to take account of this in calculations.

It was seen in chapter 3 that the bulk modulus increases with the pressure and a linear law is often a sufficient approximation.

The melting temperature also changes with the pressure and the dependence is deduced from thermodynamics. We make appeal to the Clapeyron equation which involves the latent heat of solidification L_s. If T_m is the melting temperature the entropy involved in melting is L_s/T_m. Thermodynamic arguments give the relation

$$\frac{L_s}{T_m} = \frac{dp}{dT_m} \frac{1}{\Delta V} \tag{5.30}$$

where (dT_m/dp) is the dependence of the melting temperature on the pressure. ΔV is the difference between the specific volumes of the solid and liquid phases. The use of equation (5.30) to find T_m at different pressures depends on our knowing L_s over a range of pressures, but this information is not known for the higher pressure range.

Empirical relations are valuable and one often used, relating pressure p to melting temperature T_m, is that due to Simon

$$p = A\left[\left(\frac{T_m}{T_0}\right)^c - 1\right] \tag{5.31}$$

where T_0 is the melting temperature at zero pressure and A and c are numerical constants. In some accounts c is related to the Grüneisen parameter γ:

$$c = \frac{6\gamma + 1}{6\gamma + 2}$$

Metals (with the exception of mercury and bismuth) appear to show a linear relation between T_m and volume compression ΔV. If a is a numerical constant this means

$$\frac{T_m}{T_0} = 1 + a\left(\frac{\Delta V}{V_0}\right)$$

but the rule must be regarded with caution at high pressures.

Other relationships from thermodynamics will be introduced in later chapters as they are needed.

5.14 CONCLUSIONS

1 Radioactive decay provides heating of the interior of a planetary body.

2 The equilibrium of a planetary body does not depend on the temperature, but the particular conditions inside do.

3 The distribution of heat within the body cannot be determined uniquely from surface measurements. There is uncertainty about the distribution of radioactive materials within the body and the initial temperature distribution at formation is unknown.

4 The internal transfer of heat can be by conduction or by convection. Radiation effects play no sensible part in the transfer.

5 The internal temperature distribution for a body under simple compression is adiabatic. For convective motion to occur the local temperature gradient must exceed the adiabatic value.

6 Conditions inside a planetary body quite generally are very likely to give rise to convective motion with very high viscosity. The magnitude of the Prandtl number will be very high and that of the Reynolds number very low in comparison with normal laboratory conditions.

REFERENCES AND COMMENTS

RC5.1 GENERAL TRANSFER OF HEAT

The transfer of heat is described on the basis of the conservation of energy in the form of heat. A classic text for heat conduction is

Carslaw H S and Jaeger J C 1967 *Conduction of Heat in Solids* 2nd edn (Oxford: Clarendon)

A wide range of matters are contained in this book which is an invaluable reference.

Consider a fixed volume V of material. The heat content H within the volume will change with time due to two causes: heat can pass out (or in) through the surface, and heat can be produced (or annihilated) within the volume. There are only these two possibilities. The equation of heat transport is an expression of this statement.

The heat content ΔH of a fixed volume of material will change with time due to two causes: one, ΔH_S, is the heat passing *into* the volume in unit time and the other, ΔH_P, is the heat *produced* in the volume in unit time. The conservation of heat as a form of energy is expressed by

$$\Delta H \, \mathrm{d}t = -\Delta H_S \, dt + \Delta H_P \, \mathrm{d}t$$

applying to the time interval dt. The heat flux $\boldsymbol{q}$ at any point on the surface $\boldsymbol{S}$ enclosing the volume is summed over the whole surface to obtain the total energy passing through. Then

$$H_S = \oint \boldsymbol{q} \cdot \mathrm{d}\boldsymbol{S}$$

where d$\boldsymbol{S}$ is the surface element at the particular point on $\boldsymbol{S}$. The surface integral of $\boldsymbol{q}$, being the total heat flowing through the surface, is by definition the divergence of $\boldsymbol{q}$ for the total volume (the divergence theorem), so that

$$\oint \boldsymbol{q} \cdot \mathrm{d}\boldsymbol{S} = \int_V \mathrm{div}\, \boldsymbol{q} \, \mathrm{d}V. \tag{RC5.1}$$

The production of heat in the volume can be expressed in terms of the heat produced per unit volume H_V

$$H_P = \int_V H_V \, \mathrm{d}V. \tag{RC5.2}$$

The effect of the heat conducted into and that produced in the volume is to increase the heat content of the volume. If c_p is the heat capacity per unit volume and ρ the density of the material comprising the volume we have

$$\Delta H = \int_V \rho c_p \frac{\partial T}{\partial t} \, \mathrm{d}V \tag{RC5.3}$$

where T is the temperature at the time t. Using equations (RC5.1), (RC5.2) and (RC5.3) the expression for energy conservation for the volume V becomes

$$\int_V \left(\rho c_p \frac{\partial T}{\partial t} - \mathrm{div}\, \boldsymbol{q} - H_V \right) \mathrm{d}V = 0.$$

This condition is to apply at all points within the fixed volume, irrespective of its size, and this can be so only if the integrand itself vanishes, so that

$$\rho c_p \frac{\partial T}{\partial t} = \mathrm{div}\, \boldsymbol{q} + H_V.$$

This is the general statement of the energy balance for the volume.

In applying the statement in practice it is necessary to express $\boldsymbol{q}$ in terms of the temperature. For simple heat flow, Fourier's law applies and

$$\boldsymbol{q} = -k \, \mathrm{grad}\, T.$$

The sign convention accounts for the observed fact that heat flows from the hotter to the colder regions. Heat is *lost* to the volume by conduction, but a positive sign means that heat is gained; div $\boldsymbol{q}$ must, therefore, carry a negative sign. For the loss of heat to the volume

$$-\text{div}\ \boldsymbol{q} = -\text{div}(-k\ \text{grad}\ T) = \text{div}(k\ \text{grad}\ T).$$

For a homogeneous heat conductor, k is constant and

$$\begin{aligned}\text{div}(k\ \text{grad}\ T) &= k\ \text{div}(\text{grad}\ T)\\ &= k\,\nabla^2 T\end{aligned}$$

where ∇^2 is the Laplacian. For cartesian coordinates

$$\nabla^2 = \partial^2/\partial x^2 + \partial^2/\partial y^2 + \partial^2/\partial z^2$$

while for spherical symmetrical geometry (r, θ, ϕ)

$$\nabla^2 = \frac{1}{r^2}\frac{\partial}{\partial r}\left(r^2\frac{\partial}{\partial r}\right) + \frac{1}{r^2 \sin\theta}\frac{\partial}{\partial\theta}\left(\sin\theta\frac{\partial}{\partial\theta}\right) + \frac{1}{r^2\sin^2\theta}\frac{\partial^2}{\partial\phi^2}.$$

This gives the equation

$$\rho c_p\, \partial T/\partial t = k\,\nabla^2 T + H_V \tag{RC5.4}$$

as the equation for the distribution of temperature in a homogeneous conductor. The distribution of temperature in space at a given time is linked to the time evolution of the distribution. The spacial variation enters through the second derivative while the time enters by only the first derivative; this mathematical structure is typical of irreversible processes. The initial non-uniform distribution of heat evolves into a final uniform distribution. The conservation statement (RC5.4) has the form of a diffusion equation.

The equation (RC5.4) is best put into dimensionless form. For a spherical body of radius R_P we introduce the variables

$$\tau = Kt/R_P^2 \qquad \xi = r/R_P \qquad \theta = T/T_m$$

where T_m is the melting temperature. The solution of equation (RC5.4) is sought with boundary conditions $\theta = \theta_0(x, y, z)$ initially and $\theta = \theta_s$ at the surface. We write

$$\theta = \theta_n + \theta_v$$

where θ_v is the solution of

$$\nabla^2\theta_v - \frac{1}{K}\frac{\partial\theta_v}{\partial t} = 0$$

such that $\theta_v = 0$ at the surface and initially

$$\theta_v = \theta_0 - \theta_u.$$

Then θ_u is a solution of

$$\nabla^2\theta_u = -H_V/K$$

and at the surface

$$\theta_n = \theta_s.$$

This form of solution reduces the original problem involving both a space and a time dependence to two separate problems, one of a steady temperature and the other of a variable temperature with prescribed initial general and surface temperatures. The procedure is readily extended to include radiation from the surface.

The precise form of the solution for a planetary body depends critically on the initial conditions, such as whether the accretion was hot or cold, and on the precise structure of the interior.

RC5.1.1 The approximation of Polhäusen. Although the task of solving equation (RC5.4) correctly analytically is possible in some cases of geophysical interest, the most general solutions are numerical. There is an advantage in having an approximate analytical solution available, accurate to about 10%, to act as a guide to other more accurate studies. A way of achieving this is by adapting a method first proposed by Polhäusen. The essential feature is to find a solution of equation (RC5.4) in some average form over a region of material of specified thickness. We set:

$$A = R_P^2 H_V/k \qquad \text{(RC5.5)}$$

so that equation (RC5.4) becomes

$$\nabla_\xi^2\theta = \frac{\partial\theta}{\partial\tau} - A. \qquad \text{(RC5.6)}$$

It is supposed that A is a known function of location and time; in practice this means assigning the radioactive heat content and accounting for its exponential time decay at a rate controlled by the decay constant λ.

Integrate equation (RC5.4) between the depths (usually the surface and some depth E)

$$\int_0^E \nabla_\xi^2\theta \, d\xi = \int_0^E \frac{\partial\theta}{\partial\tau} \, d\xi - \int_0^E A \, d\xi. \qquad \text{(RC5.7)}$$

The integral on the left-hand side will yield an explicit expression involving the temperature gradient (and so the heat flux) at the two depths. The second integral on the right-hand side can be evaluated since A is given. The first integral can be evaluated if the thermal gradient is known. This is not the case exactly, but an approximation can be used, for instance a linear dependence on the depth, if the range of integration is not too large. The approximation arises in the specification of the internal temperature gradient.

RC5.1.2 Cooling of a semi-infinite molten slab. To see the application of the approximation, consider a semi-infinite slab of homogeneous material initially at the melting point T_m, of known thermal properties with a free surface. The surface will radiate heat to the surroundings and will fall below the melting point: a solidification front is formed which moves inwards and we wish to estimate the rate at which it moves.

We use equation (RC5.7) integrated between the surface and the depth E of the solidification front. It is natural to use cartesian coordinates and to choose the ξ-direction vertically downwards with the origin in the plane surface. Then equation (RC5.7) becomes

$$\int_0^E \frac{\partial}{\partial \xi}\left(\frac{\partial \theta}{\partial \xi}\right) \mathrm{d}\xi = \int_0^E \frac{\partial \theta}{\partial \tau} \mathrm{d}\xi - \int_0^E A \, \mathrm{d}\xi.$$

Sunpose θ to increase linearly with depth:

$$\theta = a + b\xi.$$

Because $\theta = 1$ when $x = E$ and $\theta = \theta_s$ when $x = 0$ we have

$$a = \theta_s \qquad b = (1 - \theta_s)/E.$$

This gives

$$\frac{\partial \theta}{\partial \tau} = (1 - \xi/E) \frac{\partial \theta_s}{\partial \tau} - [(1 - \theta_s)/E^2] \frac{\partial E}{\partial \tau} \xi$$

and so

$$\int_0^E \frac{\partial \theta}{\partial \tau} \mathrm{d}\xi = \frac{1}{2} \frac{\partial \theta_s}{\partial \tau} E - \frac{1 - \theta_s}{2} \frac{\partial E}{\partial \tau}.$$

Then

$$\left(\frac{1 - \theta_s}{2}\right) \frac{\partial E}{\partial \tau} = \frac{1}{2} \frac{\partial \theta_s}{\partial \tau} E - \left.\frac{\partial \theta}{\partial \xi}\right|_E + \left.\frac{\partial \theta}{\partial \xi}\right|_0 - \int_0^E A \, \mathrm{d}\xi.$$

We have yet to account for the heat fluxes at $\xi = 0$ and $\xi = E$.

At $\xi = 0$ we are dealing with the surface heat flux associated with radiation, and so

$$\left.\frac{\partial \theta}{\partial \xi}\right|_0 = \gamma \theta_s^n$$

where n is the dimensionless radiation constant. For Stefan cooling $n = 4$ while for Newton cooling $n = 1$.

At $\xi = E$ the heat flux is supplied by the latent heat of fusion L_F as the material solidifies. Conservation of heat energy gives immediately

$$\left.\frac{\partial \theta}{\partial \xi}\right|_E = \beta_0 \frac{\mathrm{d}E}{\mathrm{d}\tau}$$

where β_0 is the Stefan number defined by

$$\beta_0 = \frac{L_F}{c_p T_F}.$$

Equation (RC5.5) now takes the form

$$\left(\beta_0 + \frac{(1-\theta_s)}{2}\right)\frac{dE}{d\tau} = \frac{1}{2}\frac{\partial\theta_s}{\partial\tau}E - \gamma\theta_s - \int_0^E A\, d\xi.$$

The simplest form of this expression is when the surface temperature is held constant and heat loss is by Newtonian cooling ($n=1$) over a sufficiently short interval of time for the radioactive heat to remain constant. Then the speed of the solidification front is easily determined by assigning θ_s and A.

Detailed application of the method to material of planetary interest gives two useful results. First, molten silicate material initially molten at the melting point will solidify to a depth of about 100 km in some 2×10^8 yr in the absence of heat sources. Second, the surface temperature falls quickly to a low value (over perhaps 10^6 years) and stays sensibly constant from then onwards. In preliminary calculations, at least, the surface temperature can be supposed to have the present value unless the very earliest periods of the history of the planet are being considered. These magnitudes could well be shortened if short-lived radioactive elements have effectively decayed. These considerations could be relevant to conditions in the outer regions of the terrestrial planets and particularly for the Moon, the Jovian and Saturnian satellites and perhaps Mercury.

For application to the deep interiors of the larger planetary bodies it may well be necessary to account for the effects of pressure on the thermal properties.

RC5.2 DIMENSIONLESS NUMBERS

The arrangement of equations into dimensionless form is the normal starting point for the analysis of a particular physical problem and this leads to the recognition of associated dimensionless groupings of physical variables, i.e. numbers. There is a range of numbers relevant to fluid mechanics, and there is a wide range of books describing fluid mechanics, but we quote two of especial interest:

Landau L D and Lifschitz E M 1958 *Fluid Mechanics* (London: Pergamon Press)

Tritton D J 1977 *Physical Fluid Dynamics* (Victoria: Van Nostrand Reinhold)

The description of fluid flows is made on the basis of the conservation of mass, momentum and energy. Because mass is conserved, the flow of fluid from a given fixed volume in space means the fluid density inside is reduced. If ρ is the local fluid density and v the local fluid velocity, the conservation of mass is expressed by

$$\rho \, \mathrm{d}\rho/\mathrm{d}t = -\mathrm{div}\, \boldsymbol{v}$$

where

$$\mathrm{d}/\mathrm{d}t = \partial/\partial t + (\boldsymbol{v} \cdot \mathrm{grad})$$

is the total derivative moving with the fluid.

The fluid velocity can be assigned the general magnitude U over the region L during the time t_0; we can then write

$$x = x'L \qquad v = v'U \qquad t = t't_0$$

where x' and v' are dimensionless and show how the speed varies at different points in the region. The density similarly can be expressed in terms of a datum magnitude ρ_0 as $\rho = \rho'\rho_0$. The equation of mass conservation (usually called the continuity equation) is then

$$\left(\frac{\rho_0}{t_0}\right)\frac{\partial \rho'}{\partial t'} + \left(\frac{U\rho_0}{L}\right)(\boldsymbol{v}' \cdot \mathrm{grad}')\rho' + \left(\frac{U}{L}\right)\rho_0\rho' \,\mathrm{div}\, \boldsymbol{v}' = 0. \qquad \text{(RC5.8)}$$

This involves several terms of the same dimension and each term is made dimensionless by division by one of the terms as standard. We choose the (second) inertia term as the standard. Dividing equation (RC5.8) throughout by $U\rho_0/L$, we find the form

$$\left(\frac{L}{Ut_0}\right)\frac{\partial \rho'}{\partial t'} + (\boldsymbol{v}' \cdot \mathrm{grad}')\rho' + \rho' \,\mathrm{div}\, v' = 0 \qquad \text{(RC5.9)}$$

The combination of variables Ut_0/L is dimensionless and is called the Strouhal number St. Then

$$(1/St)\,\partial\rho'/\partial t' + (v' \cdot \mathrm{grad}')\rho' + \rho' \,\mathrm{div}\, v' = 0 \qquad \text{(RC5.10)}$$

is the dimensionless form of the continuity equation.

The conservation of momentum in the flow is expressed for a viscous Newtonian fluid (where stress is proportional to rate of strain) by the Navier–Stokes equation. This must be written in component form with ε_{ijk} the Levi–Civita symbol. Then

$$\rho\left(\frac{\partial v_i}{\partial t} + v_j \frac{\partial v_i}{\partial x_j}\right) = \rho g_i(1 - \beta\, \Delta T_i) - \frac{\partial p}{\partial x_i} + \mu \frac{\partial^2 v_i}{\partial x_j\, \partial x_j} + 2\rho v_j w_k \varepsilon_{ijk} \qquad \text{(RC5.11)}$$

where rotation with angular velocity w is included and the pressure p includes also the rotation contribution. Thermal differences in the fluid volume are accounted for through the coefficient of thermal expansion β and the external force is supposed to be gravity with acceleration g.

We invoke a well known approximation, named after Boussinesq, according to which thermal variations of the density are to be disregarded except for the term involving gravity. This means that the density throughout the Navier–Stokes equation is to be assigned some mean value; there is no need to show this explicitly in the formula. Introducing the further datum variables g_0, μ_0, w_0, T_0, p_0 so that

$$g=g'g_0 \qquad \mu=\mu'\mu_0 \qquad w=w'w_0 \qquad T=T'T_0 \qquad p=p'p_0$$

the Navier–Stokes equation (RC5.11) becomes

$$\left(\frac{\rho_0 U}{t_0}\right)\frac{\partial v_i'}{\partial t'}+\left(\frac{\rho_0 U}{t_0}\right)\rho' v_j' \frac{\partial v_i'}{\partial x_j'}=\rho_0 g_0 \rho' g_i'-(\rho_0 g_0 \beta_0\, \Delta T)\rho' g_i' \beta'\, \Delta T_i'-$$
$$\frac{p_0}{L}\frac{\partial p'}{\partial x_i'}+\frac{\mu_0 U}{L^2}\mu' \frac{\partial^2 v_i'}{\partial x_j'\, \partial x_j'}+\rho_0 U w_0 2\rho' v_j' w_k' \varepsilon_{ijk}. \qquad \text{(RC5.12)}$$

Each term involves a combination of datum variables and a prescription to apply these to a particular point in space. To make each term dimensionless we divide throughout by the dimensionality of any one of them; let us again choose the inertia term and so divide throughout by the combination $(\rho_0 U^2/L)$. This gives

$$\left(\frac{L}{Ut_0}\right)\rho' \frac{\partial v_i'}{\partial t'}+\rho' v_i' \frac{\partial v_j'}{\partial x_j'}=\left(\frac{Lg_0}{U^2}\right)\rho' g_i'-\left(\frac{Lg_0\beta_0\, \Delta T_0}{U^2}\right)\rho' g_i'\, \Delta T_i'-$$
$$\left(\frac{p_0}{\rho_0 U^2}\right)\frac{\partial p'}{\partial x_i'}+\left(\frac{\mu_0}{p_0 LU}\right)\mu' \frac{\partial^2 v_i'}{\partial x_j'\, \partial x_j'}+\left(\frac{w_0 L}{U}\right)2\rho' v_j' w_k' \varepsilon_{ijk}. \qquad \text{(RC5.13)}$$

There are six dimensionless groupings of variables here and these lead to various numbers which specify the flow. The first combination on the left-hand side is simply the Strouhal number introduced above:

$$St=Ut_0/L. \qquad \text{(RC5.14a)}$$

The first term on the right-hand side describes the ratio of gravity to inertia forces and forms the Froude number Fr:

$$Fr=U^2/Lg_0. \qquad \text{(RC5.14b)}$$

The third accounts for the ratio of the buoyancy to inertia forces and is rearranged to extract the speed of flow

$$g_0\beta_0\, \Delta T_0 L/U^2=(\rho_0^2 g_0\beta_0\, \Delta T_0 L^3/\mu_0^2)(\mu_0^2/U^2 L^2 \rho_0^2).$$

The first bracketed expression is the Grashof number Gr

$$Gr=g_0\beta_0\, \Delta T_0 L^3/v_0^2 \qquad \text{(RC5.14c)}$$

where $v_0=\mu_0/\rho_0$ is the kinematic viscosity. The expression in the second bracket is the square of the coefficient of the fourth term on the right-hand side of the Navier–Stokes equation, accounting for the ratio of viscous and inertia forces. This term is the Reynolds number Re

$$Re = UL/\nu_0 \tag{RC5.14d}$$

The second coefficient on the right-hand side of equation (RC5.13) is then written Gr/Re^2.

The Grashof number itself is conveniently rearranged; introduce the thermometric conductivity K_0 as the datum to get

$$Gr = (g_0\beta_0\,\Delta T_0 L^3/K_0\nu_0)(K_0/\nu_0). \tag{RC5.14e}$$

The second bracket involves properties of the fluid only and not of the flow—this is the Prandtl number Pr. The first bracket involves the thermal variables and is the Rayleigh number Ra. Explicitly we have

$$Pr = \nu_0/K_0 \tag{RC5.14f}$$

$$Ra = g_0\beta_0\,\Delta T_0 L^3/K_0\nu_0. \tag{RC5.14g}$$

The third term on the right-hand side of equation (RC5.13) describes the effects of pressure; these effects are controlled by the equation of state of the fluid and we need not consider them here. The last term of the equation compares the effects of rotation and inertia; this defines the Rossby number Ro

$$Ro = U/w_0 L. \tag{RC5.14h}$$

The ratio of the rotation to viscous forces, on the other hand, is obtained through the rearrangement

$$w_0 L/U = (w_0 L/\nu_0)(\nu_0/UL).$$

The second bracket is the inverse of the Reynolds number; the first is the Taylor number Ta

$$Ta = w_0 L/\nu \tag{RC5.14i}$$

The Navier–Stokes equation (RC5.13) is now written in dimensionless form

$$\left(\frac{1}{St}\right)\frac{\partial v_i'}{\partial t'} + v_j'\frac{\partial v_i'}{\partial x_j'} = \left(\frac{1}{Fr}\right)\rho' g_i' - \frac{Gr}{(Re)^2}\,\rho' g_i'\beta'\,\Delta T' - \left(\frac{p_0}{\rho_0 U^2}\right)\frac{\partial p'}{\partial x_i'}$$
$$+\left(\frac{1}{Re}\right)\mu'\,\frac{\partial^2 v_i}{\partial x_j'\,\partial x_j'} + \left(\frac{1}{Ro}\right)2\rho' v_j' w_k' \varepsilon_{ijk} \tag{RC5.15}$$

showing explicitly the various numbers associated with the forces acting on the fluid element. This form of the equation is the most general for application to a range of flow conditions and fluids. If any two flows are such as to have all the associated numbers with the same magnitudes the flows are identical from a mathematical point of view. Although the conditions may differ widely in scale, knowledge of the conditions of one situation will provide complete information about the other. This is the basis of modelling and expresses the dynamical similitude of the flows. This subject is of very wide practical importance and can be taken further in

Birkhoff G 1955 *Hydrodynamics; a study in logic, fact and similitude* (New York: Dover)

To this point, conditions have been assumed isothermal but thermal flows are treated in the same way. The conserved quantity is now the energy, both of fluid motion and of fluid thermodynamic content. The expression of energy conservation appropriate to fluid flows is

$$\rho c_{\mathrm{p}} \frac{\partial T}{\partial t} + \rho c_{\mathrm{p}} v_j \frac{\partial T}{\partial x_j} = \lambda \frac{\partial^2 T}{\partial x_j \, \partial x_j} + H + \Phi \qquad \text{(RC5.16)}$$

where c_{p} is the specific heat capacity at constant pressure, Φ is the heat due to viscous dissipation and H is any heat produced per unit volume of the material per unit time. As before, we introduce dimensionless variables. The form of the viscous dissipation term is dependent entirely on the nature of the fluid. In this analysis, this is a simple Newtonian fluid but the materials of planetary interiors are not likely to follow such a form in detail. We will not, therefore, pursue the viscous terms at this stage. We will also neglect the heat production term. The remaining terms take the form

$$\left(\frac{L}{Ut_0}\right) \frac{\partial T'}{\partial t'} + v'_j \frac{\partial T'}{\partial x_j} = \left(\frac{K_0}{UL}\right) K' \frac{\partial^2 T'}{\partial x'_j \, \partial x'_j}. \qquad \text{(RC5.17)}$$

The new number here is the coefficient on the right-hand side, called the Peclet number Pe:

$$Pe = UL/K_0. \qquad \text{(RC5.14j)}$$

It easily follows from equations (RC5.14d) and (RC5.14f) that

$$Pe = RePr. \qquad \text{(RC5.14k)}$$

The role of the Prandtl number in specifying thermal flow conditions can be appreciated. The Peclet number describes the ratio of convected to conducted heat. For strong conduction $Pe < 1$ whereas for strong convection $Pe > 1$.

Conditions at the boundary are particularly important to scale and the Nusselt number allows this to be achieved in certain cases. Consider a surface at temperature T with the surroundings at the lower temperature T_0. The heat flow from the surface to the surroundings is written

$$q = H(T - T_0).$$

This is rearranged into the form of a conduction expression applying to the dimension L by writing

$$q = kH(T - T_0)L/Lk = Nuk(T - T_0)/L$$

where

$$Nu = HL/k \tag{RC5.141}$$

is the Nusselt number.

RC5.3 RADIATION CONTRIBUTION TO HEAT FLOW

There was, some years ago now, the thought that the usual thermal conduction of heat might be supplemented by radiation contributions even within the Earth. A radiation conductivity results which, however, has been found to make at most a negligible contribution to the overall heat flow. There is no present evidence that anything other than the ordinary thermal conductivity need be introduced to provide the discussion of thermal effects.

6

MAGNETIC FIELDS OF INTERNAL ORIGIN

Magnetic fields occur quite generally in the Solar System. Some are associated with the Sun, some with the planets and others with the regions in between. The magnitudes of the various magnetic energies are very small in comparison with other energies we have been considering and so are not involved in determining the overall equilibrium of a planetary body. Nevertheless, the study of magnetic effects of internal origin can be important for our present purposes because it allows us to make inferences about conditions inside the body and can act as a guide to the construction of models.

That the Sun is the source of a magnetic field is clear from a variety of observations (for example, the trajectories of flares, the occurrence of solar–terrestrial atmospheric phenomena and so on). Magnetic field components are carried away from the Sun by the solar wind and this is probably the origin of the main components of the interplanetary fields. Planets show magnetic effects. Some are associated with an atmosphere (where there is one) while others are associated with the central planetary condensation. The total measured field near the surface can be separated analytically into a component of internal origin and a remainder of external origin, using a mathematical technique first discovered by Gauss. It is the component of internal origin that interests us now.

6.1 OBSERVED MAGNETIC FIELDS

The Earth's magnetic field is by far the best known and studied although radio measurements have revealed a general field for Jupiter and space missions have revealed other magnetic fields including a substantial field for Saturn of internal origin.

6.1.1 THE TERRESTRIAL FIELD

The effect of the Earth's magnetic field on a small piece of lodestone suspended freely at its centre of mass was known to the ancients, who used the effect for purposes of navigation. It was William Gilbert in 1600 who associated the effect with magnetic properties of the Earth itself. His statement that the Earth is a great permanent magnet reflects the fact that the measurements available to him of the field at the surface were very closely similar to those that would be found were the Earth a uniformly magnetised sphere. The same configuration would result from a small bar magnet located near the centre of the Earth. Although Gilbert's prediction has proved to be true in a poetic sense only, the condensed Earth is certainly the source of magnetic energy and the form is dipolar to a high degree of approximation, but with the centre of symmetry displaced about 400 km from the centre of the Earth. The details of the field have been explored in ever more detail since Gilbert's time and especially so over the last three decades with the advent of space vehicles.

The magnetic field has a strength of about 0.4 gauss at the equator and about 0.6 gauss at the poles. Such strengths are found in small magnets used for picking up pins. Although the Earth's field appears to be so weak by laboratory standards it must not be concluded that its effects are entirely negligible for every situation found at the surface. The magnetic dipole moment of the Earth is 8.0×10^{23} J T^{-1} and this is not a small magnitude. The mean axis of the dipole makes an angle of about 11.5° to the rotation axis and the direction of the magnetic vector is opposite to that of the angular momentum (rotation) vector. Such a configuration is said to be a magnetically antiparallel. Whether the field is parallel or antiparallel at a particular time is to some extent accidental because palaeomagnetic measurements of rocks at the Earth's surface give evidence for reversals of the direction of the terrestrial field at various times in the past. The time between reversals is variable and we are not dealing with a simple periodic phenomenon. For instance, the field showed many reversals in the period 0–50 million years ago, but considerably fewer in the period 50–70 million years ago. Over the lifetime of the Earth neither orientation appears to have been favoured over the other. The oldest known rocks (some 3800 million years old) show evidence of palaeomagnetism: this both establishes the magnetic field of the Earth as a long-lasting feature and shows that the reversal mechanism is also a long-standing element of the field mechanism. The orientation of the axis appears to remain fixed in space but the field changes direction by collapsing and building up again, the north and south poles interchanging. The time for the reversal is very short, probably about 10^3 years. The invariant direction of the field suggests an association with a dynamically conserved quantity and the close alignment between the magnetic and rotation axes suggests that it is the rotation features of the planet that are to be linked to the magnetic field.

It has been recognised, for instance from navigation over the centuries, that the Earth's field is a dynamic entity. Both the dipole and non-dipole components change position and strength with time and the effects are very marked on maps showing the magnitudes of the inclination and declination of compass needles over the last two or three hundred years, during which time records have been kept at two or three magnetic observatories (for details see RC6). The dipole component shows a wobble around the poles, keeping the dipole form on the average, while the non-dipole component shows a westward drift.

Various estimates of the energy required to maintain the Earth's field range between 10^9 and 10^{11} watts, the uncertainty being associated with the possible efficiency of a production mechanism (see §6.4). This scale of energy is small in comparison both with that associated with the heat flow through the surface (of the order of 10^{13} W) and that released annually by earthquakes (about 10^{12} W).

6.1.2 EFFECT ON THE SOLAR WIND

The Sun is the source of a continual emission of ionised gas (plasma) composed of atoms showing the relative cosmic abundance and arising from an expansive pulsation of the solar corona. This emission is called the solar wind, and is presumably a general feature of stars.

At the distance of the Earth from the Sun the speed of the solar wind is typically 400 km s^{-1} although this figure is subject to considerable short-term variation depending on the physical state of the solar corona. The particle density in the wind is about 10^7 m^{-3} corresponding to a temperature of some 10^5 K. Because the atoms are ionised, they interact with the Earth's magnetic field to produce characteristic effects. The Sun-side lines of magnetic force are compressed by the field while those on the other side are extended to form a magnetic tail (see RC6). The structure is encased by a surface called the magnetopause. Outside this there is a region where the speed of the solar wind exceeds the local magnetic (Alfven) wave speed and here the region is marked by a bow shock on the Sunward side. This bow shock is produced by the interaction between the main magnetic field and the solar wind and the recognition of an extended shock in observations is an indicator of the existence of a magnetic field associated with the planet itself. The extent and nature of the bow shock is determined by the strength and geometry of the magnetic field. Care must be taken in the interpretation of observations, however, because perturbations of the flow of the solar wind can also be brought about by effects arising from ionised layers in an atmosphere quite separately from the presence of a main magnetic field. These effects will generally be very small but need not be smaller than those associated with a very weak main field.

It seems then that further evidence is needed to distinguish the presence of a general magnetic field of internal origin and this can be obtained from the particular distribution of charged particles in the ionised layers of the atmosphere. The general field will trap charged particles for periods of time and for the Earth these form the Van Allen radiation belts. Polarised radiation will be emitted by the trapped particles, the details of which will depend on the field strength and orientation.

6.1.3 OTHER TERRESTRIAL PLANETS

Space missions to other planets have shown the existence of such bow shock structures there. Their presence cannot, however, be taken as complete evidence for the existence of a general magnetic field of the planet.

The first extraterrestrial planetary object studied in detail was the Moon (by a succession of American and Russian fly-bys, orbiters and landers). It was found early on that the Moon does not possess a general magnetic field within the limits of measurement and that any magnetic moment must be smaller than the Earth's by a factor of at least 10^{-7}. There is no magnetosphere or bow shock and the solar wind strikes the surface directly except for a few days each month around the time of full moon when the Moon is in the magnetosphere of the Earth and so is hidden from the direct action of the Sun. The anti-solar side of the Moon is protected from the solar wind and a pointed plasma umbra is present encased in a plasma penumbra. The umbra region is a void which is probably the most perfect vacuum in the inner reaches of the Solar System.

Palaeomagnetism has been discovered in lunar rocks brought back to Earth with local field strengths in the range 6–300 gammas (1 gamma $= 10^{-9}$ tesla). It is not yet possible to relate these data to a general lunar configuration although there are indications that a general dipole form is possible. The age determinations of the rocks involved suggest that the initial magnetic field giving rise to the palaeofield permeated the Moon some 3×10^9 years ago although the origin of the field is still the subject of controversy. The fluctuating form of the solar wind gives rise to transient electric fields in the main lunar body and the *in situ* measurements of these currents by Apollo astronauts has allowed some information about the electrical state of the lunar interior to be deduced. As a general summary, the electrical phenomena suggest the interaction between the solar wind plasma and an inert non-magnetic but slightly conducting blunt body. In particular, there seems to be no general magnetic field maintained now.

The magnetic conditions on Mars have proved difficult to assess in spite of the presence of two landers. The magnetic moment is certainly not in excess of 4×10^{-3} that of the Earth and is probably less. The thin atmosphere is electrically conducting and the interaction with the solar wind produces a weak bow shock and possibly a magnetopause.

Mercury provided a surprise during the three fly-bys of Mariner 10 during 1973 and 1974. In spite of the planet being devoid of an atmosphere, a bow shock was found together with transient electrical effects associated with the presence of plasma. The observations are consistent with the existence of a fully developed magnetic field of internal origin and with a strength about 4×10^{-5} that of the Earth.

The magnetic state of Venus is now becoming reliably known. American and Russian spacecraft have observed the planet with increasing sophistication since 1962 and it is now clear that there is no general magnetic field with a strength in excess of 5×10^{-5} that of the Earth. The dense atmosphere has ionising layers (like those of the Earth) and a strong bow shock is present on the Sun-side due to atmospheric magnetism. There are no trapped particles, however, and no equivalent to the terrestrial Van Allen regions.

6.1.4 THE MAJOR PLANETS

That Jupiter is the source of a general magnetic field was known before the advent of space measurements. Radio astronomers had detected radiation at deca- and decimetre wavelengths and the decametre emissions (at a frequency of 22.2 MHz) are precisely similar to the synchrotron radiation to be expected from a system of relativistic charged particles trapped in a mean dipole field. Put another way, Jupiter has Van Allen belts. The direction of polarisation of the radiation gives an indication of the orientation of the field axis which was judged to be about 10° away from the rotation axis; this is closely the same as for the Earth. The decimetre radiation (in the range 300–3000 MHz) was found to originate in a toroidal region surrounding Jupiter again tilted at an angle of about 10° to the planet's equatorial plane.

These Earth-based observations were confirmed and extended during the encounters of the planet by Pioneer 10 and 11 when the great magnitude of the magnetic field became apparent. The magnetic moment is 1.9×10^4 times that of the Earth, giving a magnetic moment of 1.5×10^{27} J T^{-1}. The magnetic moment vector makes an angle of about 9.5° to the rotation axis and the magnetic and rotation vectors have the same direction (parallel field). The period of rotation of magnetic effects is about 9 h 55.495 min, which is to be compared with the period 9 h 50.5 min for surface markings near the equator and the period 9 h 40.68 min for higher latitude markings.

Jupiter is surrounded by an enormous magnetosphere; while it resembles that of the Earth in shape its extent is some 1.2×10^3 times as great. This comes about because the planetary magnetic energy is considerably stronger than for the Earth while the strength of the solar wind is only some 4% that at the distance of the Earth. The full extent of the Jovian magnetospheric tail has not been observed but it is likely to be very extensive and has certainly been observed as far away as the distance of Saturn. The role of the major satellites, and particularly Io, as a source of ionised particles in the Jovian environment will not concern us now.

The magnetic conditions of Saturn were quite unknown before the Pioneer missions. A bow shock was observed by Pioneer 11 at 24 Saturn radii on the

Sunward side and this was found later to be associated with a dipole field with magnetic moment about 4.4×10^{25} J T^{-1} or about 550 times that for the Earth. The direction of the magnetic vector makes an angle of less than 1° with the rotation axis and the two directions can be supposed to be coincident to a good first approximation. The magnetic and rotation vectors have the same direction, as for Jupiter but unlike the Earth. The rotation period for magnetic effects is likely to be about 10 h 39.9 min. This can be compared with the period 10 h 13 min for equatorial markings and 10 h 40 min for the rest. As for Jupiter, the fluid nature of the planet is very apparent from these times since neither planet rotates as a rigid body.

The magnetic conditions of Uranus and Neptune are unknown. There is a suggestion that observed L-α ultraviolet emission from the polar regions of Uranus could arise from auroral effects in atmospheric hydrogen and, if this proved to be the case, these observations would suggest strongly the presence of a dipole field on the planet. There will be considerable interest in the magnetic data obtained by Pioneer 11 when it reaches Uranus in 1986. Although there is a common expectation that a substantial magnetic field will be found both for Uranus and for Neptune this need not be the case because the behaviour of Jupiter and Saturn could well be a poor indicator of the behaviour of the other major planets.

6.2 ORIGIN OF MAGNETIC FIELDS

It would seem natural for all planetary magnetic fields to arise from a single physical mechanism, although the details could well differ from one planetary type to another. For instance, the detailed structure of a terrestrial-type planet differs markedly from that of a major planet. While these differences can be included in a single physical mechanism the source of necessary energy will be determined by local circumstances and this could mask to some extent the single mechanism applying to all cases.

It is not part of our present aim to study the current ideas on the production of planetary magnetic fields in detail but only to see in what way the presence or absence of a magnetic field can allow information about the internal conditions to be deduced. It is, however, useful to give some account of the theoretical work to act as a background to our more general discussion.

The original statement by Gilbert that the Earth behaves as a great permanent magnet cannot be anything other than superficial appearance. The temperature increases with depth in the Earth and the Curie temperature of the material (in the range 700–1000 K for most minerals of interest) is very quickly passed, the thin surface crust being unable to support an inherent magnetisation of sufficient strength to provide the observed field. The discovery that the field is not entirely dipolar in geometry and also shows continual, though small, changes in time even including field reversals adds weight to the initial conclusion that simple magnetism cannot be the explanation. An alternative description in terms of an array of electric currents due to various maintained potential differences in the solid Earth is no more satisfactory. Nor is the hypothesis that magnetism is in some way a natural accompaniment of rotation.

The same arguments would seem to be applicable immediately to the other terrestrial planets. With the exception of the association of magnetism with rotation, none of the mechanisms for producing a magnetic field considered so far appear applicable to the major (fluid) planets because solid material would seem necessary if the observed intensity of magnetisation is to be achieved. Another mechanism is necessary if it is to apply generally throughout the Solar System.

The common feature of all the members is a fluid interior probably showing good electrical conductivity in some regions although the specific physical conditions of the liquid interior will differ from one planetary body to another.

A flowing fluid which is also a conductor of electricity (and we remember here particularly the ferrous components of the cosmic abundance) can have an associated magnetic field and the energy for the field now derives from mechanical sources and not from differences of electrical potential. The material can withstand stronger mechanical than electrostatic forces and the association of an external magnetic field with internal material motions can be sustained on physical grounds even for the largest bodies. The conversion of mechanical into magnetic energy can be described generally as a dynamo mechanism and is, in fact, the inverse of the mechanism used to provide commercial electrical power. This comparison must not be interpreted too literally and the associated topological requirements will be entirely different.

It must be admitted at the outset that this dynamo mechanism was first proposed and considered because all alternative origins for the magnetic field had failed to provide a plausible account. The development of the theory must, as a consequence, involve a close blending between theoretical expectations of an unusual situation and the features of observed planetary magnetic fields.

To produce a magnetic field the fluid pattern must change sufficiently rapidly with time. The shortest time scales will be measured in years and this will imply that the viscosity of the region where the source of the magnetic field is located must be low enough to respond quickly to the controlling external forces. The source of the field must, in this way, be associated with a planetary region having genuinely fluid properties in that the resistance to shear must be low. In this respect, the magnetic field characteristics differ markedly from the thermal ones, where the time scale is long even in terms of the age of the Solar System itself. For the terrestrial-type planets (see chapter 8) this will be in the central core region; for the major planets (see chapter 7) the source can be more generally distributed.

6.3 GENERAL REQUIREMENTS OF A DYNAMO MECHANISM

There are three general requirements for a successful dynamo theory of the planetary magnetic field. First, there must be a mechanism for the conversion of energy (probably that of kinetic motion of the fluid) into magnetic energy. Second, the energy requirement must be low enough to be within the overall evolutionary budget of the planet. This will set limits on the efficiency required of the basic mechanism. Third, the mechanism must be stable and self-sustaining. There is the supplementary requirement that whatever form the conversion into magnetic energy will take, the external field must have the overall appearance of a dipole field to an appropriate approximation. This requirement might need to be met by introducing an intermediate coupling between the internal (presumably toroidal) and external (dipolar) fields. This would imply that the field we see outside is in some sense a secondary phenomenon, a suggestion fully compatible with the observed reversals of the Earth's field over geological history.

The magnetic field emanating from a source can be separated into a succession of components associated with dipole, quadrupole and higher multipole characteristics. The amplitude of these components will decrease with distance from the source, the dipole form decaying the slowest. Any field distribution assumes more nearly the dipole form the farther it is viewed from the source. This means that an observed field of closely dipole form can be supposed to be further from the source than one with dipole–quadrupole characteristics. Such an observation can have practical value; for instance, the external field of Jupiter has a stronger multipole character than that of the Earth. This could imply a terrestrial source deeper in the Earth than is the corresponding source for Jupiter.

6.4 DETAILS OF THE MECHANISM

We can best approach the discussion of planetary magnetic field production by first outlining the overall mechanism and then setting down the mathematical statement for its action.

6.4.1 *THE MAGNETOHYDRODYNAMIC COUPLING*

The basic idea for the dynamo mechanism arises from magneto-hydrodynamics (MHD). Suppose an electrically conducting fluid is moving relative to a permeating and already existing magnetic field of external origin. The motion of the fluid constitutes an electric current which will interact with the magnetic field according to the laws of electricity. A force (the Lorentz force) $\boldsymbol{F}_L$ will act between the current $\boldsymbol{j}$ and the field $\boldsymbol{B}_0$:

$$\boldsymbol{F}_L = \boldsymbol{j} \times \boldsymbol{B}_0 \quad (6.1)$$

with a magnitude proportional to the product jB_0 and a direction perpendicular both to that of the current and the field. The effect of this force is to change the motion of the fluid. At the same time, the motion of the fluid (being equivalent to an electric current) will have a magnetic field $\boldsymbol{b}$ associated with it which augments the external field to become $\boldsymbol{B}=\boldsymbol{B}_0+\boldsymbol{b}$. It is this augmented field that now affects the new motion of the fluid which again acts to augment the magnetic field controlling its motion. A mutual interaction exists between the fluid motion and the field and it is tempting to envisage an equilibrium in which a balance of energy exists between the fluid motion and the field. Such a condition, in which energy is equipartitioned between the kinetic and magnetic forms, will have the form $\rho v^2/2 \sim B^2/4\pi$, relating the field to the flow. This simple mechanism will allow the conversion from kinetic to magnetic energies up to the field strength where the two forms of energy have comparable magnitude and so will satisfy the first requirement for the theory set down in §6.3. It is necessary to begin with a seed magnetic field but this can be arbitrarily small. The MHD mechanism is capable of magnifying the initial field provided the flow pattern is appropriately chosen. What the internal flow pattern is to be for a given external field is the heart of the problem. It is also the case that the final field need bear no resemblance to the initial field; the production process eliminates the initial conditions.

6.4.2 *THE MATHEMATICAL EQUATIONS*

What we have been describing is a link between the equations of the hydrodynamics of a conducting fluid and the electromagnetic equations of Maxwell. The controlling equations for the energy balance are obtained by merging these two sets of equations together. The linking physical variable is the electrical conductivity of the material comprising the fluid.

The electromagnetic equations of Maxwell summarise the electric contributions of the flow. We write the electric vector $\boldsymbol{E}$, the electric induction $\boldsymbol{D}$, the magnetic field $\boldsymbol{B}$, the magnetic induction $\boldsymbol{H}$ and the electric current $\boldsymbol{j}$. The flow is subject to the approximation that the displacement current is negligible, so the equations become

$$\text{curl}\,\boldsymbol{H}=\boldsymbol{j} \qquad \text{curl}\,\boldsymbol{E}=-\partial\boldsymbol{H}/\partial t$$

$$\boldsymbol{D}=\varepsilon\boldsymbol{E} \qquad \boldsymbol{H}=\mu\boldsymbol{B}. \tag{6.2}$$

This approximation implies no net electric charge accumulation and since there are supposed to be no free poles we have also

$$\text{div}\,\boldsymbol{E}=0 \qquad \text{div}\,\boldsymbol{B}=0. \tag{6.3}$$

The electric current in the material is written in terms of the flow velocity $\boldsymbol{v}$ and the magnetic field by the vector multiplication

$$\boldsymbol{j}=\sigma[\boldsymbol{E}+(\boldsymbol{v}\times\boldsymbol{H})]. \tag{6.4}$$

An equation for calculating the magnetic field strength and direction at any point is obtained by combining equations (6.2), (6.3) and (6.4). First take the curl operation of each component of equation (6.2) and the curl operation of equation (6.4) and combine the two results. This gives the vector equation

$$\frac{\partial \boldsymbol{B}}{\partial t}=\mathrm{curl}(\boldsymbol{v}\times\boldsymbol{B})+\frac{1}{\sigma\mu}\nabla^2\boldsymbol{B} \tag{6.5}$$

as the means of calculating the magnetic field strength in the fluid.

The two terms on the right-hand side have different physical interpretations. If the electrical conductivity is indefinitely large only the first term remains and so

$$\partial\boldsymbol{B}/\partial t=\mathrm{curl}(\boldsymbol{v}\times\boldsymbol{B}). \tag{6.6}$$

The magnetic field is frozen into the fluid because (according to equation (6.4)) any relative motion between the fluid and the field would give rise to electric currents of infinite magnitude, which is not permissible. The magnetic field is carried along by the fluid and equation (6.6) refers to a direct convection of the field by the fluid.

The other extreme is when the fluid has a finite electrical conductivity and is at rest. Then equation (6.5) takes the form

$$\frac{\partial \boldsymbol{B}}{\partial t}=\frac{1}{\sigma\mu}\nabla^2\boldsymbol{B}. \tag{6.7}$$

This is of the form of a diffusion equation and represents the tendency of the field to diffuse out of a region in this case. Only when a solid conductor is infinitely conducting will the magnetic field remain of constant magnitude in any volume without the support of an external source of energy.

The full expression (6.5) contains a contribution from both these limiting cases. Whether the magnetic field at any point will grow or diminish depends on the relative effectiveness of the convection and diffusion mechanisms. This, for its part, depends on the nature of the flow field and the magnitude of σ. Here is the basic mechanism for magnetic field amplification. As to orders of magnitude, for a fluid flow with characteristic speed U in a volume of characteristic dimension L^3, the field $\boldsymbol{B}$ will be amplified according to equation (6.5) if the right-hand side is positive, that is if $UB/L>B/\mu\sigma L^2$ or if $\mu\sigma UL>1$. The physical significance is clearer if we refer to equation (6.7). This diffusion equation has the same form as that for the vorticity ($=\mathrm{curl}\ \boldsymbol{v}$) and $1/\mu\sigma$ plays the same role as the kinematic viscosity. The magnetic equivalent of the Reynolds number $Re=UL/\nu$ is the magnetic Reynolds number $Re_{\mathrm{m}}=\mu\sigma UL$, and the most primitive condition for field amplication is that $Re_{\mathrm{m}}>1$. If this condition is not fulfilled the magnetic field will decay. The field amplification is the more marked the larger σ, and the condition of frozen-in field corresponds to $Re_m \to \infty$. This condition must be avoided in practice for two reasons: first, there would be no possibility of field amplification for such a frozen-in condition, and second, there would be no possibility of the field escaping from the region to be observed outside. While it is necessary that $Re_m > 1$, it should not be too great. It seems likely from detailed theoretical considerations that $Re_m \sim 10$ is probably a good condition for dynamo action.

The remaining problem of providing the appropriate fluid flow pattern involves the hydrodynamic part of the problem. The fluid flows under the action of various forces according to the requirements of the conservation of mass, momentum and energy. One force is the Lorentz force, as described in §6.4.1 and this involves both the flow velocity and the field vector. The same two quantities appear in equation (6.5) and the two sets of mathematical equations are therefore linked.

The energy equation has a trivial form if the flow is isothermal and the continuity equation reduces to the simple statement div $\boldsymbol{v} = 0$ for incompressible flows. The statement of momentum conservation remains and this is the Navier–Stokes equation for a Newtonian fluid. The mathematical form of this equation is strongly non-linear and few exact solutions have yet been found.

We can approach the matter in terms of the comparison of the effectiveness of the different forces, i.e. using the dimensionless numbers already used in chapter 5 and considered in RC5. There is a new number to introduce now which arises from the presence of the Lorentz force. Comparing the strengths of the Lorentz and inertia forces, respectively F_L and F_I, we find the new ratio $\sigma UB^2L/\rho U^2 = \sigma BL/\rho U$ to be important. Remembering the magnetic Reynolds number introduced above, we have

$$F_L/F_I \sim Re_m(\mu B^2/\rho U^2). \tag{6.8}$$

The quantity in the bracket is a dimensionless number called the Alfven number $Al = \mu B^2/\sigma U^2$. We can therefore write $F_L/F_I = Re_m Al$.

The Reynolds number Re and magnetic Reynolds number Re_m, together with the Alfven number and the Taylor number (related to the Rossby number accounting for rotation effects and defined in equation (RC5.14h)) are the relevant dimensionless groupings for the dynamo problems. The essential feature is a Reynolds number that is not too small (associated with high fluidity) and rotation. The rotational characteristics are important because the fluid will be contained within a restricted (albeit large by laboratory standards) region and the flow can only take place if it continually circulates (convects) within the region. This implies a rotational motion, although the geometry may not be simple. The source of energy to drive the flow has yet to be specified.

A rotating fluid has an angular momentum which will be coupled to the angular momentum of the planet as a whole. This will include a component due to whole body rotation about the axis. Requirements of conservation of angular momentum will mean that the many separate rotational features of individual small-scale fluid eddies will be coupled to the overall planetary rotation. A relation between the mean observed magnetic axis of symmetry and the planetary rotation axis is, therefore, to be expected.

The field production then has two parts: one is the direct stochastic motion of the electrically conducting fluid directly involved with the conversion of mechanical into magnetic energies, and the other is the coupling effect which produces the mean poloidal field which is observed outside. The stochastic field is likely to have a toroidal form due to the shearing action of the overall rotation.

These arguments have been developed in a formal way by many authors, although it has not yet proved possible to provide a full model able to account for the details of the observed field of any planetary body. There is general agreement, however, that this mechanism is the correct one in principle and that the full specification of fields actually observed will ultimately be accomplished.

6.4.3 LIMITATIONS OF SYMMETRY

There is a limit to the degree of symmetry to be encountered in a steady dynamo mechanism, as was shown by T G Cowling in 1934. In particular, a magnetic field symmetrical about an axis cannot be maintained by symmetric motion. The deduction of this important result involves the result of the competition between the diffusive and convective aspects of the magnetic field changes. Because the electrical conductivity is finite, a magnetic field in a volume of fluid will decay, and because the electrical conductivity is not zero, fluid motion will convect lines of force. For symmetric motion, the fluid transports field lines bodily in a representative plane through the axis without creating new lines of force. With no new field being produced, there is no source of field to balance the diffusive loss and a steady state of the field is not possible.

The argument can be expressed simply in mathematical terms. A poloidal (dipole) field consists of closed lines of magnetic force centred on a neutral ring (see figure 6.1) around the axis of the field. The field can be regarded as maintained by an electric current flow around the ring, the direction of the field being determined by the direction of the current. If the current stops, the field collapses into the ring, which is a neutral ring for the system. It follows from equation (6.2) that curl $\boldsymbol{B}$ is not zero in the ring (because there is an electric current flowing there) but the magnetic field $\boldsymbol{B}$ is zero there. The electric current cannot, therefore, be maintained by a fluid motion through the electric force according to equation (6.4) because this involves $(\boldsymbol{v} \times \boldsymbol{B})$, but $\boldsymbol{B}$ is zero. The topology of the field about an axis of symmetry also means that the current cannot be maintained by electrostatic forces. It follows that the steady maintenance of the field is not possible.

This theorem had a great influence on the development of the theory

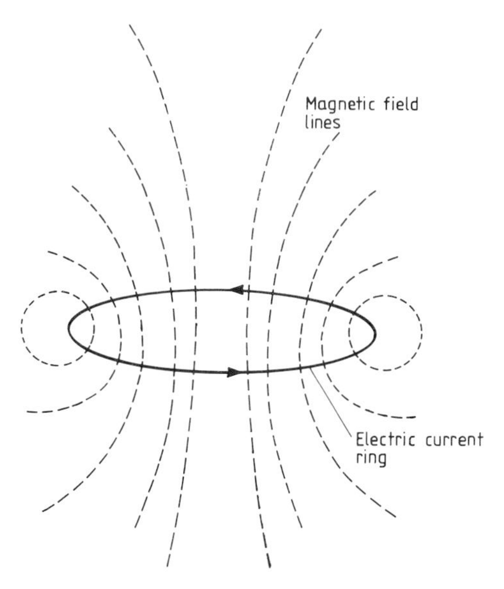

Figure 6.1 *A poloidal magnetic field.*

because it has the form of an anti-dynamo theorem. In particular, the observed symmetry of the dipole field presented difficulties of understanding in the light of the theorem. The importance of the theorem must not be either over- or underestimated. There is no reason to suppose that the observed field is the primary dynamo field (indeed, polarity reversal of the terrestrial field would suggest this is not so). Any observed symmetry in the external field could well be associated with a coupling between the primary dynamo field and the poloidal field observed outside, and this coupling is not involved in the Cowling theorem.

6.4.4 EXISTENCE ARGUMENTS

Although our concern here is not with a detailed account of the development of the dynamo theories it is interesting to notice the consequence of the publication of the Cowling theorem. With any easy solution of the problem precluded, emphasis was placed on the search for existence theorems and particular circumstances (peculiar from a planetary point of view) where a magnetic field was undeniably maintained. This involved some laboratory studies and computational work as well as analytic approaches. The first existence theorem was due to Herzenberg who, in 1958, showed that a dynamo mechanism is possible as the result of the interaction between two indefinitely small rotating electrically conducting spheres embedded in an electrically conducting material with which they are each in full electrical contact. The mechanism involves the twisting of field lines and the associated distance scales allow the preferential decay of small-scale components of magnetic field in comparison with the large-scale components. A second approach, also in 1958, was reported by Bachus for a sporadic dynamo which could be regarded as steady only in the mean. A short period of vigorous fluid convection is to be followed by a longer period of rest during which again preferential decay of small-scale components can take place. These approaches are important not for their application to planetary problems but because they show the dynamo mechanism can work in circumstances that can be properly defined.

A third approach (Childress, 1967, 1970; Roberts, 1970; see Moffatt, 1978) considers periodic dynamos. It appears that a field in the form of a travelling wave can be maintained by steady cellular three-dimensional convection, periodic in each spatial direction.

More recent work (Parker, Braganskii, Steenbeck and Krause, Moffatt (see Moffatt 1978)) has been based on the mean-field electrodynamics of a rotating sphere. The motions are supposed to be nearly axisymmetric and the consequences of the Cowling theorem are avoided by arranging for some mean azimuthal current to arise which prevents the poloidal field from shrinking down to the neutral ring. It seems that some form of Belltrami flow (where curl $\boldsymbol{B}=\text{constant}\times\boldsymbol{B}$) is involved showing the importance of helicity in the theory. The concentration on mean-field approaches can well be of special significance in the future, because it is fully consistent with an underlying fluid, turbulent, convective motion (the associated Rayleigh number is usually very high); unfortunately the methods for treating such systems are little understood at the present time.

6.5 SOURCE OF MAGNETIC FIELD ENERGY

The energy associated with maintaining the Earth's field has already been quoted as in the range 10^9–10^{11} W. This is not a large amount of energy to find even if the conversion mechanism from kinetic to magnetic energy is as low as 1% overall. Mechanisms suggested to provide this level of energy have centred on some element of stirring. One possibility is that the elliptic orbit of the Moon round the Earth, linked to the bulges and distortions of the figures of the Earth and Moon, will cause such a stirring effect on the fluid inner region of the Earth. Another such convection effect will arise from the solidification of the core. If the central fluid is solidifying at the outside and the solid portions are falling inwards, a mixing will occur, the solid particles releasing energy by falling through the gravitational field. This potential energy is then available for conversion to magnetic energy.

To gain some estimate of the energy involved we remember that the mass of the core is about 2×10^{24} kg and that the radius of the outer core is very closely double that of the inner core. This means (since the densities are comparable) that the mass of the outer core is about 7 times that of the inner core, or the mass of the outer core is about 1.5×10^{24} kg. If this quantity of material were to solidify over 10^{11} years this would be 1.5×10^{13} kg per year. There are 3.15×10^{7} seconds in a year so the corresponding solidification rate for the mass of the outer core is $1.5 \times 10^{13}/3.15 \times 10^{7} = 4.8 \times 10^{5}$ kg s^{-1}. The energy E_g released by this mass falling through the gravitational field is about

$$E_g = 4.8 \times 10^5 \times 3.5 \times 1.5 \times 10^3 \text{ J} = 2.52 \times 10^9 \text{ J}$$

where we have taken the acceleration of gravity to be 3.5 m s^{-2} and the depth of the present core layer to be 1.5×10^3 km. This energy is at the lower end of that required to maintain the present terrestrial magnetic field. There is also the latent heat of solidification L_s available for use. Taking $L_s = 2.6 \times 10^4$ J kg^{-1}, the solidification of the total core would provide the additional energy $E_s = 1.2 \times 10^{10}$ J s^{-1}, or rather more than that due to phase separation. This amount of energy would be adequate to drive an efficient dynamo and so to provide the present observed field strength. The energy released from solidification would not necessarily provide convection; the falling of the solid material under gravity will. The overall mechanism can, therefore, be envisaged in which the supply of energy arises from the solidification process whereas the conditions for the magnetic dynamo come from the consequent falling of solid material to accumulate at the centre of the Earth. The quantity of energy available in the past from both phase separation and solidification will have been rather higher than that calculated here, while that available in the future will be less because the conditions in the core are now well defined. If these mechanisms apply and if the efficiency of the conversion of kinetic to magnetic energy is not too low the present magnitude of the Earth's field can be understood in terms of solidification energy.

The alternative convective mechanism involving thermal convection alone is not so easily applied in this case. The temperature gradient in the core is small (and will vanish at the centre) while gravity there is also small. The amount of energy available is therefore rather less than that involved with a solidification/separation process. On the other hand, thermal convection can be expected to be important in the major planets and could act as a dynamo source, as we shall see in chapter 7.

6.6 CONSEQUENCES FOR STUDIES OF THE INTERIOR

On the basis of the arguments developed so far we see that the observation of an external magnetic field can allow a great deal to be inferred about interior conditions. To begin with, it implies the existence of an electrically conducting fluid region of sufficient scale to allow a convective motion to be set up. Second, it implies a source of energy. For a silicate body of terrestrial type this can arise from some form of phase change mechanism (solidification or liquefaction). For a fluid body, such as Jupiter or Saturn, the energy could arise from the gravitational separation of components, such as helium separating from hydrogen. Thirdly, some form of rotation is implied. No one of these contributions alone will be adequate and the existence of the field suggests the simultaneous operation of them all. The absence of a magnetic field in a particular case will not, as a consequence, generally be ascribable to one simple cause although the obvious absence of one component will vitiate the others and make the production of a magnetic field impossible.

6.7 LOCATION OF THE SOURCE OF THE FIELD

An interesting possibility for locating the conducting region associated with the magnetic field source has been suggested by Hide. Changes of the external field are due primarily to changes in the source region. One cause will be the convective motion in the region, which will be transmitted to the external field lines because the internal field will be partially frozen into the conducting material. For changes over a short period of time (less than the diffusive decay time of the field), the lines will be largely locked into the fluid and therefore not able to move separately. By following back the observed changes of the lines of force into the planetary interior a level will be reached where the field configuration does not change. This will be the outer boundary of the electrically conducting fluid. Preliminary calculations along these lines have led to the recognition of a boundary for the Earth consistent with that known from seismic studies and for Jupiter at a depth of some 0.8 of the radius, which is consistent with the deduced magnetic dynamics based on independent arguments (see chapter 7).

6.8 CONCLUSIONS

1 Planetary magnetic fields of internal origin arise from the conversion of kinetic energy of material motion into magnetic energy.

2 The theory of the production of a planetary magnetic field (dynamo theory) is based on a combination of the separate theories of electromagnetism (Maxwell's equations, excluding the displacement current) and hydrodynamics (explicitly the Navier–Stokes equation).

3 The link made through the Lorentz force and the electrical conductivity of the material is a critical parameter. While its magnitude must be high enough for a sufficient coupling to result it must not be so high that the magnetic field becomes frozen into the fluid. It is likely that the magnetic Reynolds number should be greater than 10 but not by more than an order of magnitude or so.

4 The source of energy for maintaining the field will involve a convection of matter within the body. This could be caused by differentiation or solidification of material, or it could arise from thermal motions.

5 The particular manifestation of the mechanism in particular cases will be dictated by the physical characteristics of the volume concerned and considerable accidental variation of the production can be expected for different planetary bodies. The details for a terrestrial planet will therefore differ from those for a major planet.

6 The field observed outside the body is likely to be a mean secondary effect linked to the production region by a separate coupling mechanism. Field reversals will be associated with the coupling.

7 The integrated effect which is observed outside will be affected by the full body rotation of the planet. The Rossby number can be used to describe this influence.

8 The theory has not yet been worked out in detail for a particular planetary body and our understanding of the processes is still far from complete.

9 The observation of a planetary magnetic field of internal origin is a potential probe to conditions inside the body. The quantitative development of the theory will ultimately provide a valuable additional tool for exploring interiors.

REFERENCES AND COMMENTS

RC6.1 OBSERVED MAGNETIC FIELDS

For details of solar properties see

Kiepenheuer K O 1959 *The Sun* (Ann Arbor: University of Michigan Press)
Meadows A J 1970 *Early Solar Physics* (Oxford: Pergamon)
Eddy J A 1979 *A New Sun: The Results from Skylab* NASA, SP-402

For details of the solar wind:

Akasofu S-I and Chapman S 1972 *Solar Terrestrial Physics* (London: Oxford University Press)
Brandt J C 1970 *Introduction to the Solar Wind* (Reading: Freeman)
Hundhausen A J 1972 *Coronal Expansion and the Solar Wind* (Berlin: Springer)

The method of separating a magnetic field of internal origin from one of external origin was first devised by Gauss. The method is based on the representation of the field as a power series of distance involving spherical harmonic functions (see RC4), the internal and external fields being separated by being associated with electric current systems of different form. The surface separating the two regions is supposed to be without electric currents. This idealised situation is not fully true in practice but is sufficiently near the actual form to act as a very good zero approximation for calculations. It is by such methods that the magnitude of the dipole component is determined and, by repeating the analysis for different epochs, the change of the dipole with time can be deduced. As a rough guide, magnetic fields of internal origin change on time scales of 10^2 years or longer, while those of atmospheric origin have shorter time scales for change.

The methods and background for the subject generally, including the specification of the field, and with many maps, is given by

Chapman S and Bartels J 1951 *Geomagnetism* (London: Oxford University Press)

This is an old book but still gives an excellent background to the field even now.

For a wide ranging, though less detailed, study see

Garland G D 1979 *Introduction to Geophysics* (Philadelphia: Saunders)

This book also contains an account of palaeomagnetism. See also

Nagata T 1961 *Rock Magnetism* (Tokyo: Marozen)

Details of the magnetospheric features of the planets generally are contained in

Beatty J K, O'Leary B and Chaikin A 1981 *The New Solar System* (London: Cambridge University Press)

Details of the results of various space missions have been published in *Science*.

The palaeomagnetic field of the Moon presents interesting questions. One possibility is that it represents a field that has ceased to function; the dynamo has stopped. This presupposes that the Moon had a core of size adequate to support a dynamo mechanism at some time in the past. The maximum possible size of the present core is a sphere at the centre of some 450 km radius—this is by no means certain (more seismic work is needed) nor is it clear that this is large enough to support a dynamo mechanism. Certainly the central region appears unable to support S-waves and so is presumably broadly fluid–plastic. It is not necessary for the composition to be iron, only sufficiently electrically conducting. There are other possibilities. It could be that the lunar volume was permeated by a magnetic field of external origin (from the Earth?) in the past and the present field is a fossilised remnant of earlier electric currents. It may be that the field observed now is simply the result of bombardment in the past. The striking of the surface of the Moon (or any other body for that matter) causes the immediate neighbourhood to melt and on solidification the rock will take on a palaeomagnetism in sympathy with any local field at the time. This could be the terrestrial field if, as is possible, the

Earth–Moon distance were smaller then than now. The palaeomagnetic field of the Moon is real and its origin remains still to be elucidated.

RC6.2 DETAILS OF THE DYNAMO MECHANISM

A comprehensive study including the modern work is

Moffatt H K 1978 *Magnetic Field Generation in Electrically Conducting Fluids* (London: Cambridge University Press)

This book contains many references, particularly for the newer developments involving the mean field. Another useful reference is

Rädler K-H 1981 Mean-Field Theories of Planetary Magnetism *Adv. Space Sci.* **1** 219–29

This gives a very readable account of the mean field approach. The non-linear behaviour is investigated by

Krause F and Roberts P H 1981 Strange Attractor Character of Large-scale Non-linear Dynamos *Adv. Space Sci.* **1** 231–40

The idea behind the mean field approach is to overcome the Cowling theorem. This is done by supposing the magnetic field to be composed of a steady part and a fluctuating part. Maxwell's equations are to hold in the mean only and are to be averaged suitably over space or time to provide mean relationships between variables. The term $(\boldsymbol{v} \times \boldsymbol{B})$ in the Lorentz force then introduces a mean electric field E' into the magnetohydrodynamic equations. This fluctuating electric field is linear both in the fluctuating magnetic field and its space gradient, and is able, for the appropriate choice of geometry, to provide the azimuthal components not associated with an axisymmetric distribution. Modifications and generalisations of the approach, based on arguments of tensor symmetry, provide additional parameters which allow the observed external field to be represented with greater precision. The approach shows promise for the future but has not yet passed beyond the preliminary stages. It is likely, however, that any initial phenomenological success with the approach will ultimately lead to a deeper understanding of the basic mechanism of magnetic field production.

Since this account was written a highly interesting and readable review has been published by

Stevenson D J 1983 Planetary Magnetic Fields *Rep. Prog. Phys.* **46** 555–620

This gives a wide coverage of the present ideas relating to the dynamo mechanism and considers the possible application to essentially every member of the Solar System and the likelihood of a dynamo mechanism operating there is assessed in each case. There is a comprehensive reference list and the reader is referred particularly to this review.

7

PLANET OF LARGE MASS

The theoretical discussion of chapter 2 shows a maximum mass and size for a planetary body and Jupiter comes close to reproducing this situation in practice. There are three other very large planets (Saturn, Uranus and Neptune) and we now compare the properties we might expect such large objects to have on general grounds with those actually observed.

7.1 GENERAL CONSIDERATIONS

It is natural to begin any discussion of the composition of a planetary body by making an appeal to the cosmic abundance of the elements (see chapter 1). Because we are dealing with massive objects, the most likely major constituents are the most abundant elements, which are hydrogen and helium. Other elements will be present in only relatively small amounts. A planetary body of mass 10^{27} kg composed of the cosmic abundance would have a hydrogen/helium content of about 9.8×10^{26} kg and the remaining material would account for about 2×10^{25} kg (or rather more than 3 Earth masses). Differentiation would allow this more rare material to accumulate in a central core. If it had a radioactive heat production rate of broadly terrestrial magnitude, the core would produce heat at the rate of perhaps 8×10^{15} J s^{-1} and this would be associated with a mean temperature $\langle T \rangle \sim 4 \times 10^3$ K for the body. The temperature due to heat of accretion (see §2.8) would, of course, supplement this estimate. The total radius of such a body would be about 5×10^7 m and the overall density rather less than 10^3 kg m^{-3}. The central pressure then is of order 10^{12} N m^{-2}, which is comparable to the estimate given in RC2.6. The central density could be as high as 5×10^4 kg m^{-3}. The inertia factor is about 0.3.

This description of the gross properties does not fit the observations of the major planets in detail where it can be compared (see tables 1.1 and 1.5), although it is sufficiently close to be interesting. The overall density is too low for Jupiter and too high for Saturn; it is too low for Neptune although not too much below that of Uranus. It is significant that the calculated inertia factor for each planet other than Jupiter is too high. It seems, indeed, that the central condensation in Saturn is very strong (see table 1.5) and is nearly as strong in Uranus. That for Jupiter is weaker and for Neptune is weaker still. Each planet, it seems, must be assigned a heavy core if the inertia factors are to be understood at all. The degree of compression is lowest for Saturn and this dictates the low density there; it is worth noticing that the arguments of chapter 2 led to a density of 460 kg m^{-3} for a hydrogen sphere of maximum size and this shows that even Saturn, with a density half as large again, contains at least a small proportion of components heavier than the lightest element.

The conclusion that the main constituents of Saturn are hydrogen and helium is unavoidable from the magnitude of the mean density and it is natural to suppose that the same conclusion is to be drawn for the other members of the quartet. The only other materials that could be present with any abundance are water, methane and ammonia, but the amounts would be relatively small. The small magnitudes of the inertia factors for the bodies would seem to make it unavoidable, however, to suppose the heavier elements to make a more substantial contribution to the structure of the central regions than would be expected from the cosmic abundance alone. There is an element of doubt in that the arguments have assumed an equilibrium configuration has been achieved but this need not be the case, at least for Jupiter and Saturn. The large size of these bodies makes it possible for them still to be in the final phase of contraction.

Although each planet is unique in certain properties, Jupiter and Saturn are probably sufficiently alike to be regarded as similar and the same seems true for Uranus and Neptune. It is possible, however, that the greater similarity of Jupiter–Neptune and Saturn–Uranus on the criteria of density and inertia factor could be significant. This alternative linking together displays a slightly greater degree of symmetry of mass and distance from the Sun than that based on simple distance alone with one linked pair being encased in another. These matters could become clearer when data from the Voyager space probes now approaching Uranus and later Neptune eventually send back data.

For the present we have little option but to explore the major planets on the basis of a substantial hydrogen/helium content without assuming a strict cosmic abundance for the other components and this will be the point of view taken now.

7.2 THE PROPERTIES OF HYDROGEN AND HELIUM

There is still considerable uncertainty about the equations of state for hydrogen and helium. This is disappointing because these two materials are composed of the simplest atoms and success here is essential before more complicated materials can be approached in detail.

The range of gross physical states available to these materials seems to be agreed even though not all have been investigated experimentally. There are obviously the molecular fluid (liquid and gas) and solid divisions, but at the high pressures ($>10^{11}$ N m^{-2}) to be expected in the deep planetary interiors there is also the expectation that metallic states will be formed which may themselves be either liquid or solid. The uncertainties arise in the particular conditions of pressure and temperature needed to effect the transference from one form to the other. Most of the theoretical calculations apply rigorously to zero temperature only and it is the extension to finite temperatures that causes difficulty.

For molecular hydrogen the relation between the melting temperature T_m and the melting pressure p_m has the form

$$T_m = T_{m0}(1 + p_m/a)^{1/c}$$

where $a = 2.74 \times 10^7$ N m^{-2}, $c = 1.747$ and $T_{m0} = 14.15$ K. The Lindemann rule could then be used to extend predictions to higher pressures but the accuracy of the method is difficult to assess.

Solid hydrogen has an electronic structure which suggests an affiliation with the alkali metals and this is supposed in practice. As the pressure is raised, the compression of hydrogen reaches a level where the atomic spatial arrangement becomes more ordered. At a high pressure, variously estimated theoretically to lie in the range 0.75–2.3×10^{11} N m^{-2} and at a temperature of about 10^4 K, hydrogen is expected to take on a metallic form. The density will then be of order 10^4 kg m^{-3}. This can be either liquid or solid depending on the temperature. The ionised atoms now share electrons collectively and the ions are spatially orientated either as a liquid (with random local ordering) or as a solid (where the ordering is long-ranged). On the basis of the configuration of the atomic electrons it can be expected that metallic hydrogen will behave like an alkali metal. The possibility has been considered that even a pairing of electrons in the metal is possible to give superconducting properties. This aspect is very hypothetical but is based on the Bardeen–Cooper–Schrieffer theory. The formula deduced by these authors involves the density of Fermi states and the lattice interaction energy, together with the Debye temperature. This last quantity is unknown but could be as high as 3.5×10^3 K for metallic hydrogen. This is greater by a factor 10^2 than for alkali metals. It is this feature that allows the possibility of superconduction in hydrogen. A good survey of this whole area has been given by Cook (RC7).

The transport properties are hardly better known. One estimate of the thermal conductivity for metallic hydrogen is in the range 1–9×10^3 W m^{-1}, appropriate to conditions in Jupiter, the magnitude increasing with depth. Using the standard formulae (such as the Wiedemann–Franz law if it applies) the electrical conductivity can be estimated to have the general magnitude for metallic hydrogen 2×10^7 S m^{-1}. These estimates are certainly rough and must not be regarded as other than an initial guide for model studies.

Helium is expected to follow broadly the same features, although the transition conditions will be different. Judging from the atomic electron structure the solid form is most likely to behave as a divalent material. Of particular interest is a mixture of hydrogen and helium. Here again there is uncertainty. The most detailed theoretical forecasts are probably those of Smoluchowski. Hydrogen and helium can be mixed in a simple way; un-ionised helium will not be miscible with metallic hydrogen and so will separate out from it. At higher pressures, however, the situation could be more complicated. Once helium becomes ionised it behaves, as we have seen, as a divalent metal. Hydrogen behaves like a monovalent metal and an alloy between them can be expected which mirrors the behaviour of ordinary divalent and alkali metals. The effect of increasing pressure on the mixture is very relevant for the giant planets. At lower pressures, hydrogen and helium can mix, solid metallic hydrogen containing undissolved helium. As the pressure increases further the metallic hydrogen will form a liquid solution with helium and at the highest pressures the helium will solidify to form a solid solution of helium in hydrogen metal. The mixture will presumably alloy at the highest pressures when the hydrogen and helium atoms can be spatially arranged together.

Data for the equations of state for molecular and metallic hydrogen must still be treated with caution, but that of de Marcus probably shows the general features reliably. These are given in table 7.1. The companion data for helium are collected in table 7.2 which is also due to de Marcus and again applies at $T = 0$ K.

Table 7.1 *Pressure–density data for molecular and metallic hydrogen at 0 K, according to calculations by de Marcus.*

Pressure (10^{11} N m^{-2})	Density (kg m^{-3})		Pressure (10^{11} N m^{-2})	Density (kg m^{-3})	
	Molecular	Metallic		Molecular	Metallic
0	89		2.50	819	
0.05	238		3.00	903	1160
0.10	274		4.00		1290
0.50	428		5.00		1400
0.80	503	780	10.00		1870
1.00	548	820	15.00		2240
1.50	643		20.00		2550
2.00	732	1010	30.00		3090

Table 7.2 *The relation between pressure and density for helium at 0 K taken from calculations reported by de Marcus.*

Pressure (10^{11} N m^{-2})	Density (kg m^{-3})	Pressure (10^{11} N m^{-2})	Density (kg m^{-3})	Pressure (10^{11} N m^{-2})	Density (kg m^{-3})
2×10^{-4}	234	2×10^{-1}	1150	10.0	4560
1×10^{-3}	323	4×10^{-1}	1400	20.0	6210
1×10^{-2}	534	1.0	1890	30.0	7580
2×10^{-2}	626	2.0	2450	40.0	8640
8×10^{-2}	884	4.0	3160	80.0	11410
10^{-1}	936	8.0	4150		

7.3 ELEMENTARY MODELS

In the absence of firm information about the physical properties of the expected main constituents of the major planets there is little alternative but to invoke empirical rules. It can be expected that the expansion of the bulk modulus in powers of the pressure (see chapter 3) will still be valid for the condensed material and that the Murnaghan equation of state (3.20) will remain valid as a result. It could be necessary to proceed to the quadratic approximation (3.21) for the highest pressures. The linear form will obviously be exploited until it proves inadequate.

Hubbard has stressed the validity of the polytrope $B=2$ for the conditions of the major planets and has used the dependence $p=A\rho^2$ in this connection, where A is a factor related to the bulk modulus. This dependence has a considerable history and is often called the Laplace approximation. Its use gives the equations of the theory (such as the mass equation or the Emden equation treated in chapter 3) a linear form and so simplifies the calculations considerably. The internal structure of Jupiter, calculated on this assertion using the mass equation (3.20), is shown in figure 7.1. The inertia factor of 0.265 for the model is consistent with the arguments of chapter 4. There is no account of a core of heavier material, the planet being supposed to be of homogeneous composition throughout. This model can, incidentally, also provide data for Neptune which compare well with observation, but cannot account for the observed properties of Saturn and Uranus; the difficulty is the particularly small inertia factor for these two planets. There are two difficulties for Jupiter as well. To begin with, the model does not reproduce the measured features of the external gravitational field adequately; and the central density is merely 2.7 times the mean density (some 3.6×10^3 kg m^{-3}) which must be too low by a factor 10 according to the arguments of chapter 2. In addition, such a low density is not consistent with the formation of a solid or metallic

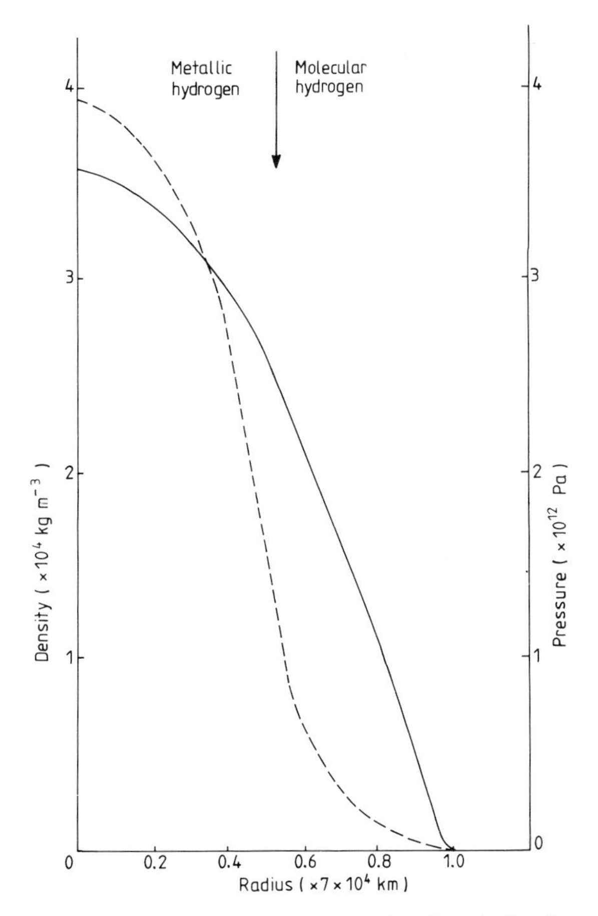

Figure 7.1 *The dependence of density on radius for a body of mass 10^{27} kg, composed of hydrogen according to the mass equation for a Murnaghan material (chapter 3) with* B = *3. Data for the pressure are included (broken line) and the possible position of the boundary between molecular and metallic hydrogen is indicated. The inertia factor for the model is* $\alpha_P = 0.265$.

phase which is necessary if a dynamo mechanism is to be available for the production of the observed extensive magnetic field. It is possible, of course, that the dynamo mechanism will act in a dense fluid (as opposed to a liquid) for which the magnetic Reynolds number is in the appropriate range (see chapter 6). The situation would be more extreme for Saturn where the central density could be as low as 2×10^3 kg m^{-3}. We can conclude again from this that a simple homogeneous model is unable to explain the observed features. This is particularly true for the predicted Saturnian gravitational field where the magnitudes of J_2 and J_4 determined from this model do not agree even in general terms with the corresponding values obtained from observation.

7.4 EMPIRICAL MODELS FOR JUPITER AND SATURN

The alternative approach is to construct models with particular observed features in mind. According to the arguments of chapter 2, the central conditions of Jupiter will be extreme in that the pressure will be high. The two criteria of particular importance are the inertia factor and the structure of the external gravitational field through the spherical harmonic expansion.

7.4.1 PRESENCE OF A CORE

The small inertia factors for Jupiter (and Neptune) imply a substantial central core of heavy material; the even smaller factors for Saturn (and Uranus) imply even stronger central condensations for these planets. Each model must, therefore, involve a core–mantle boundary at some appropriate distance from the centre. Encasing this differentiated region is a mantle of hydrogen and helium, solid and probably metallic at the base and gaseous at the top. Helium will be distributed to provide the degree of non-homogeneity required to account for the observed gravitational field distribution. The non-dipole components are stronger for both Jupiter and Saturn than for the Earth (see table 7.1) and are stronger for Saturn than for Jupiter. This suggests different distributions of material in the two cases. Models constructed on this basis are shown in figures 7.2 and 7.3. It is seen that the core size for Jupiter is some $\frac{1}{5}$ the total radius, containing rather more than five Earth masses. The corresponding core size for Saturn is rather less, corresponding to a smaller mass.

The degree of separation of hydrogen and helium may itself be greater in Jupiter than in Saturn. The movement of helium downwards through the gravitational field of the planet will release energy and it is possible that the greater outward surface energy flow of Jupiter than of Saturn (relative to that received from the Sun) arises from this source. If this is so, it would suggest a difference in the present evolutionary conditions of the two planets. The energy output from Jupiter could be met, alternatively, by a slow contraction of the planet, by about 1 mm per year, which would be expected if the planet is still in the final stage of condensation. For Saturn, the rate of contraction would need to be rather higher to provide the measured excess outflow of heat. The low mean density of Saturn can be accounted for by the small level of compression in the planet.

7.4.2 MAGNETIC CONDITIONS

The planetary body will be expected to be the source of a magnetic field according to the arguments of chapter 6 and the interior structure must allow this possibility. The presence of a general Jovian magnetic field of essentially dipole form was detected, before space vehicles were available, from

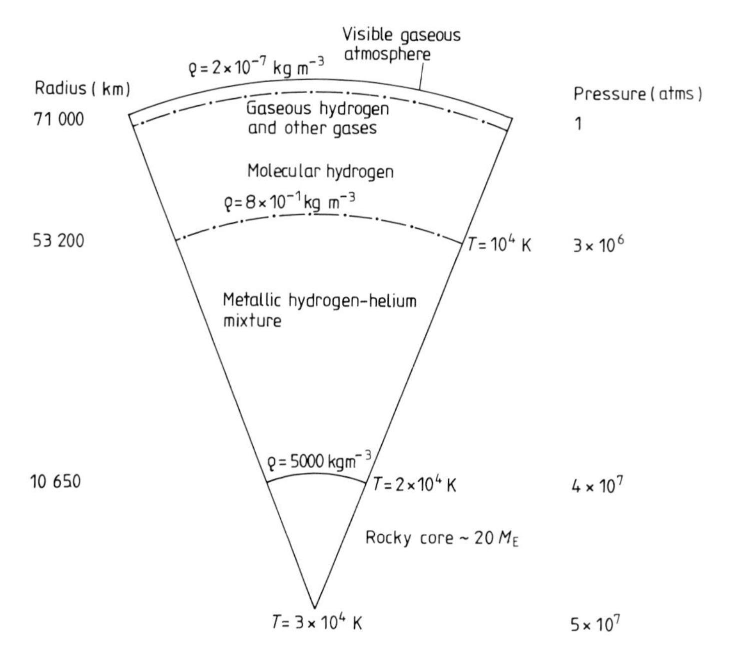

Figure 7.2 *Schematic cross section for a realistic model of Jupiter, including indications of density, pressure and temperature.*

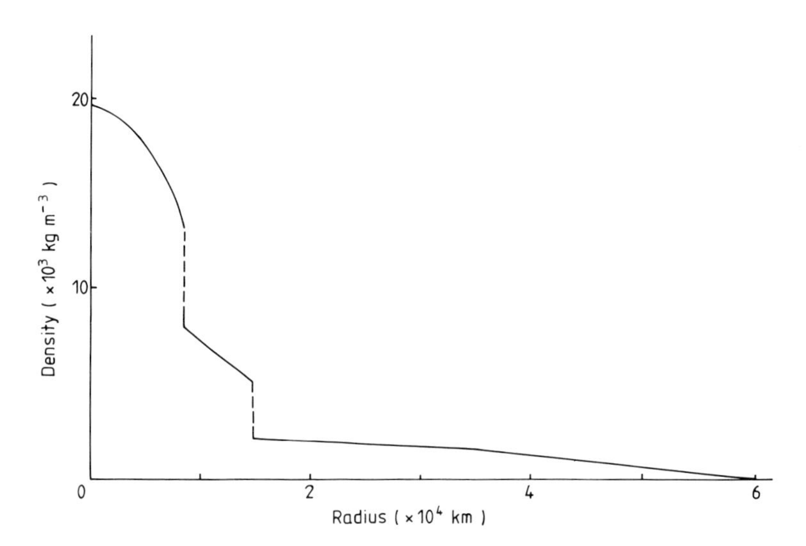

Figure 7.3 *Density against radius for a model of Saturn with a central core (Murnaghan material with* B = *3.5) and a hydrogen covering* (B = *2). The inertia factor* $\alpha_P = 0.23$.

observations of the non-thermal emission at radio (decimetre) wavelengths. This radiation is emitted by relativistic electrons trapped in the enclosing dipole field. The radiation is about 25% linearly polarised with a direction making an angle of about 9.5% with the equator. This implies a magnetic axis making the same angle to the rotation axis. The direction of the field is that of the rotation vector (parallel field, opposite to the terrestrial condition) and the strength is 10^4 times that of the Earth. The field has been extensively surveyed in the recent Voyager missions, and the associated magnetosphere detected and penetrated. Because the magnetic field lines in the surface region of the source cannot change appreciably over a short interval of time, 'plotting back' the lines at two times separated by a short interval allows the source region to be located, as the surface where the field line configuration does not change, as was pointed out by Hide (see chapter 6). Applied to Jupiter, this method suggests a source region at a depth of about 20 000 km although the accuracy is not yet high.

The magnetic dipole moment of Saturn, discovered by Pioneer 11 in 1979, has a strength some 10^3 times that of the Earth's with a dipole axis virtually in line with the rotation axis. The magnetic and rotation vectors are parallel and directed in the same way as for Jupiter and unlike the Earth; this is a coincidental arrangement (see chapter 6). This close alignment does not violate Cowling's theorem (see chapter 6) because it refers to the outer dipole field (where Cowling's theorem need not apply) and not to the internal toroidal field (where it does).

The possible modes of free oscillations of the major planets have been investigated theoretically and the appropriate frequency spectra obtained for various models. Once observational data become available for such motion more information can be expected to accrue of the interior conditions. There is little reliable information at the present time and the area is open for investigation once the accuracy of measurements is increased.

7.5 URANUS AND NEPTUNE

Our knowledge of these two planets is very rudimentary and will probably remain so until the Voyager fly-bys take place. Models for the interior must involve some heavier components than for Jupiter and Saturn because the densities are greater. The inferred inertia factors can be met only if the central regions involve a core of heavier elements. The mantle itself may also contain elements other than hydrogen and helium. The next most abundant elements (see table 1.2) are oxygen, carbon and nitrogen and compounds of these with hydrogen can be expected. Prominent among these will be water, methane and ammonia, together with various other compounds of these elements. One important observation involves the different thermal conditions of the two planets; while Uranus gives virtually no thermal emission. Neptune provides more than would be expected on the basis of primordial heating alone.

The situation is far from clear and the models proposed so far reflect these uncertainties. The simplest two-layer models involve a solid core and a liquid mantle, but are unable to provide a magnetic field source because the electrical conductivity must be too low. More recently, three-layer models have been considered consisting of a rock/ferrous core (occupying the central third of the volume), and an ice mantle (water-ice 57%, ammonia-ice 32% and methane-ice 11%) also occupies about one third of the radius; the whole is enclosed by a deep atmosphere of hydrogen and helium. Models for these two planets which best fit our present expectations are shown in figures 7.4 and 7.5. There are now regions where the conductivity will be high and magnetic effects could well result. On this basis, both Uranus and Neptune could be expected to have magnetic fields. Although the correctness, or not, of these considerations is unknown, the analyses are useful in giving information about the possible composition of large planets.

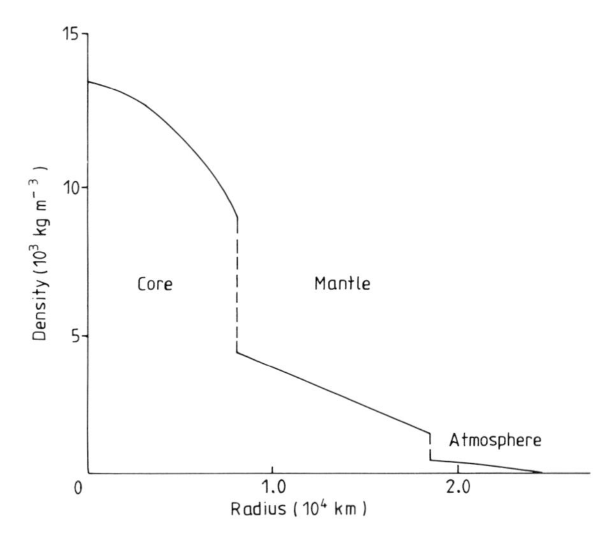

Figure 7.4 *Three region model for Uranus, with a central core (Murnaghan material with* B = *3.5), an ice mantle* (B = *3*) *and an atmosphere/crust* (B = *2*).

No direct magnetic activity has been detected from Earth-based laboratories but this need not be the end of the matter; magnetic effects of Saturn are not discernable from Earth. It is possible that an auroral-type L-α hydrogen emission has been detected from Uranus, and if substantiated this would give indirect evidence for the existence of a general magnetic field for the planet. The overall size of the major planets and the internal constitutions

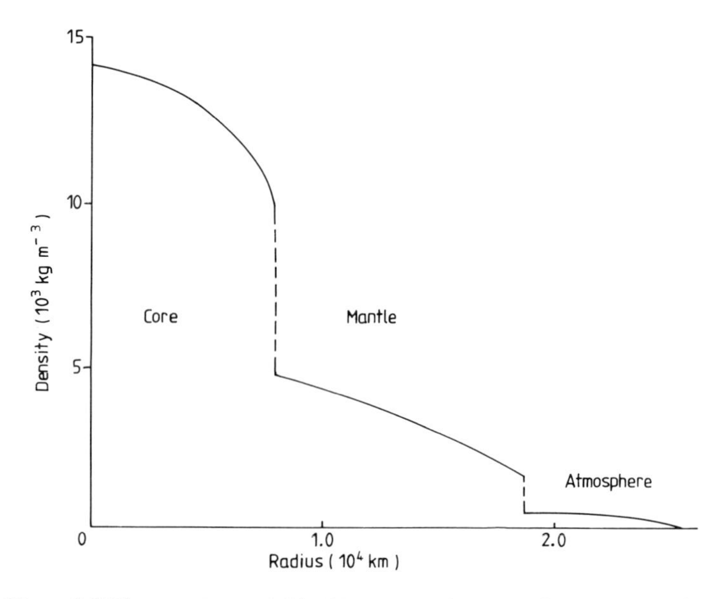

Figure 7.5 *Three region model for Neptune, with a central core (Murnaghan material with* B = *3.5), an ice mantle* (B = *3) and an atmosphere/crust* (B = *2).*

allows the possibility of the existence of general magnetic fields of internal origin and the Voyager information (expected during 1986 for Uranus and 1995 for Neptune) is awaited with some excitement. A field associated with Uranus would be particularly interesting because, presuming it lines up generally with the rotation axis which is *in* the plane of the ecliptic, the magnetic axis will have periods when it is in line with the Sun. The effects on the solar wind in the vicinity of Uranus will raise interesting problems when eventually data become available. Firmer information generally must await more refined information. The present situation is most intriguing and we can look forward to interesting developments in the future.

7.6 CONCLUSIONS

1 The composition of the largest planetary bodies is mainly hydrogen and helium but is not closely in accord with the cosmic abundance of the elements.

2 The small inertia factors for these planets imply substantial cores of heavier elements.

3 The physical properties of hydrogen and helium are sufficiently well known for many purposes of planetary studies at the present time and

equations of state are available. This involves a metallic phase for hydrogen.

4 Models of the major planets can be constructed which reproduce many of the observed properties with reasonable precision, on the basis of the known physical properties of materials.

5 The most successful models involve a central core of silicate/ferrous (rocky) materials with a mass several times that of the Earth. This is encased in a mantle of hydrogen/helium and there may be an intermediate region of water/methane/ammonia ices in Saturn, Uranus and Neptune. The hydrogen in the deep interior of Jupiter, and probably of Saturn also, will be in metallic form. This is unlikely to be the case in Uranus and Neptune where the hydrogen will be in molecular form.

6 Uranus and Neptune are likely to involve water as a particularly prominent constituent.

7 Magnetic field data for Jupiter and Saturn can be understood in general terms on the basis of these models, the sources of the magnetic energy being located in each case in a highly conducting region deep in the mantles. It is still not known whether Uranus and Neptune have magnetic fields of internal origin. This is yet another question that it is hoped will be solved by the forthcoming Voyager fly-bys of these planets. The answer will provide valuable information about aspects of the interior structures.

REFERENCES AND COMMENTS

The study of Jupiter and Saturn has been given an enormous boost by the fly-bys of the Voyager spacecraft. The literature is now very extensive and we give here a few references which will allow the reader to enter the full literature.

Close up photographs of the surface have allowed a greater understanding of the various processes involved and new features have been found. The ring system of Saturn is now realised not to be unique; Jupiter has a small ring, probably of transitory nature and in some form of dynamic equilibrium, associated with the volcanic conditions on Io. It had been found from occulation measurements that Uranus also has a ring system and there will be great interest in this connection when the Voyager craft reaches it later in this decade. Of particular interest is the confirmation and extension of previous ideas on the effects of rotation on the behaviour of Jupiter and Saturn: see

Hide R 1980 Jupiter and Saturn: Giant magnetic rotating fluid planets *The Observatory* **100** 182–93

—— 1983 The Giant Planets *Holweck Prize Lecture* The Institute of Physics (in the press)

The most comprehensive collection of analyses of Jupiter is probably

Gehrels T ed. 1976 *Jupiter: studies of interior, atmosphere, magnetosphere and satellites* (Tucson: University of Arizona Press)

A companion volume for Saturn is

Gehrels T ed. 1983 *Saturn: studies of interior, atmosphere, magnetosphere and satellites* (Tucson: University of Arizona Press)

The equation of state data for hydrogen and helium are discussed in

Cook A H 1980 *Interiors of the Planets* (London: Cambridge University Press)

where there are many references (see particularly chapter 7), and

Smoluchowski R 1967 Internal Structure and Energy Emission of Jupiter *Nature* **215** 691–5
—— 1973 Dynamics of the Jovian Interior *Astrophys. J.* **185** L95–9
—— 1981 Interiors of the Giant Planets: Recent Advances *Adv. Space Res.* **1** 103–16
Slattery W L and Hubbard W B 1976 Thermodynamics of a Solar mixture of molecular hydrogen and helium at high pressure *Icarus* **29** 187–92

The earlier calculations were

de Marcus W C 1958 *Astron. J.* **63** 2
Stewart J W 1956 *J. Phys. Chem. Solids* **1** 146

The problem of fast rotation has been considered by Trubitsyn and Zharkhov in a series of papers; see for instance

Zharkhov V N and Trubitsyn V P 1975 Fifth Approximation System of Equations for the Theory of Figure *Astron. Zh.* **52** 599–614 (Engl. Transl. *Sov. Astron. A.J.* **19** 366–72

For the modes of free oscillation see

Vorontsov S V and Zharkov V N 1981 Free oscillations of the Giant Planets *Adv. Space Res.* **1** 189–94

The possibilities of observing such oscillatory motions must depend on better data than are available at the present time. This could come about when orbiting satellites are placed round Jupiter and the fine details of the gravitational and magnetic fields are known.

For the magnetic field analysis leading to the location of the source see

Hide R 1981 The magnetic flux linkage of a moving medium: A theorem and geophysical applications *J. Geophys. Res.* **86** 681–7

The recent calculations of the interiors of Uranus and Neptune are summarised by Smoluchowski (1981): see above.

The fly-bys of these planets by Voyager II in 1986 for Uranus and 1995 for Neptune will, it is hoped, provide data to refine such models.

8

TERRESTRIAL-TYPE PLANET

The Earth is the standard for describing the four (or five if the Moon is included) planets nearest the Sun and called collectively the terrestrial planets. All have now been visited by space vehicles (either by fly-by, remote landing or, in the case of the Moon, by manned landing) and the Earth itself has been studied in great detail by a range of techniques. The extensive information which has resulted is now available to act as a guide to the study of the other members of the group. The difficulties of studying planetary interiors is shown by the Earth; although we walk and live on it we are still lacking direct and detailed knowledge of conditions inside.

The terrestrial planets are characterised by high density (in excess of $3000\ \mathrm{kg\,m^{-3}}$) and so are composed of silicate and ferrous materials which are associated with the rarer components of the cosmic abundance (see table 1.2). They have radii less than 7×10^3 km and masses less than 7×10^{24} kg and so are physically small in comparison with the fluid major planets considered in the last chapter. Unlike the major planets, the terrestrial planets contain hydrogen and helium in trace amounts only. They are to be regarded as solid from an everyday point of view, which means that their interior dynamics are controlled generally by extremely high viscosities.

8.1 SOME GENERAL CONSIDERATIONS

The interior of such a planetary body will be expected to be differentiated, with the free ferrous components (possibly containing some lighter materials such as sulphur) forming a core which underlies the silicates which constitute the mantle. The precise details of the structure cannot be predicted without a knowledge of the composition of the material which initially condensed to form the body and of the chemical processes which occurred subsequently.

The interior structure to be expected of a planetary body of mass 10^{24} kg and composed of silicate and ferrous materials, but with a composition compatible with the cosmic abundance, can be constructed from the data contained in table 1.2. The hydrogen/helium components must first be disregarded which means losing a mass of about 4.9×10^{25} kg, or about 50 planetary masses. Once this is done, the ferrous component will have a mass of 8×10^{22} kg while the non-ferrous components have a mass of 9.92×10^{23}. The total radius of the body will be about 3.5×10^3 km if the overall density is 5000 kg m^{-3}.

If all the ferrous material has differentiated to form a core (an unlikely circumstance because some iron silicates can be expected to form during differentiation) the core will contain about 8% of the total mass. Under the conditions of compression at the centre of such a body, the radius of the core would be about 1.2×10^3 km which is about $\frac{1}{3}$ the total radius. The proportion of ferrous core material deduced for the terrestrial planets is larger than this (see §8.3), indicating that they are not composed of material distributed according to even a subset of components of the elementary cosmic abundance.

8.2 APPLICATION TO THE EARTH

The theory of elastic waves as the basis for the analysis of the trajectories of earthquake (seismic) disturbances has been applied to the study of the Earth's interior over the last hundred years with ever more refinement. Today, the technique is able to yield a very detailed picture of the interior not unlike that resulting from an x-ray scan, if this were possible (see RC8 for some details).

8.2.1 GENERAL STRUCTURE

The analysis is dominated by the overall result that whereas P-waves with speed V_P can propagate throughout the entire volume of the Earth, S-waves with speed V_S are excluded from entering a central spherical shell enclosed by the outer surface of radius $R_c = 0.545 R_E$ (where R_E is the mean radius of the Earth) and the inner surface of radius $0.198 R_E$. They move readily in the encasing mantle. This result is interpreted as showing that the overall structure is that of a solid mantle (being the region where both P- and S-waves propagate) enclosing a liquid outer core region with essentially no resistance to shear, which itself encloses a solid inner core. The details of the way in which this picture has been built up (see figure 8.1) is highly interesting and can be followed from the references listed in RC8.

Apart from the division into mantle and core, the seismic waves distinguish

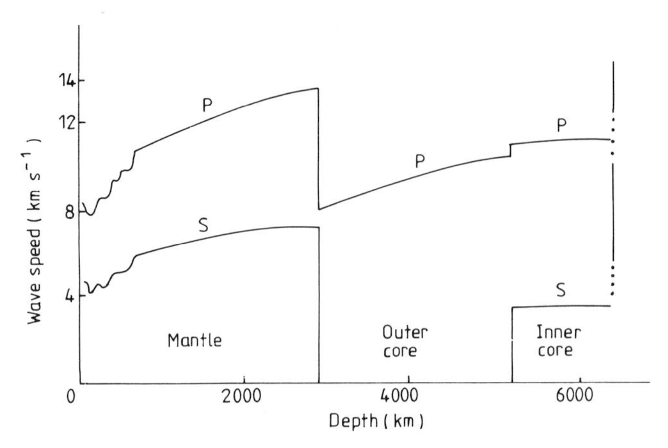

Figure 8.1 *Dependence of seismic P-wave and S-wave speeds on depth within the Earth.*

a surface layer (the crust) separated from the mantle by a sharp discontinuity (the Mohorovičić discontinuity) at a depth of 10 or so kilometres under continental regions and 2 or 3 kilometres under oceanic regions. The crust has turned out to be a complex region much affected by the thermal conditions determined by the heat flow from below and the heat of radioactive decay within. The thin crust can be regarded as a thermal boundary layer for the Earth. This overall structure arises from very general considerations which can be expected to apply to each of the terrestrial planets, although the details will differ from one planet to another. Quite generally, we must expect a condensed planet of the terrestrial type to have this basic crust–mantle–core structure.

8.2.2 DENSITY AND PRESSURE

The general results outlined above are converted into numerical magnitudes by using the arguments of the previous chapters. The starting data are the inferred magnitudes of V_P and V_S as a function of the depth within the Earth, resulting from measurements of the earthquake disturbances at the surface stations. According to equation (RC8.1) these data lead immediately to the magnitude of K/ρ as a function of the depth according to

$$K = \rho(V_P - 4V_S/3). \tag{8.1}$$

If the physical nature of the material comprising the Earth were known so that K were known numerically with depth then equation (8.1) would allow the density to be deduced at each depth. This is not the case, however, and further arguments are necessary.

The material is in hydrostatic equilibrium below a shallow depth (usually

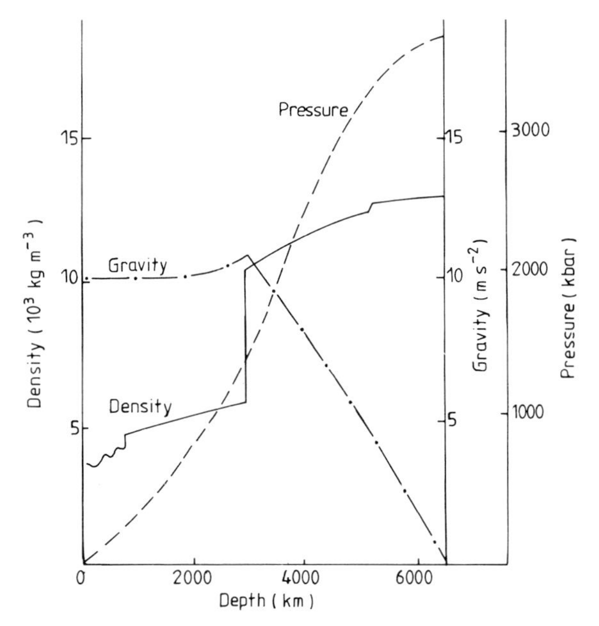

Figure 8.2 *Variation of density, gravity and pressure with depth in the Earth, deduced from seismic data using the Adams–Williamson approach.*

taken to be 33 km) and the analysis of §3.3 is then available to allow density and pressure to be deduced as a function of the depth. The results of these studies for the Earth are collected in figure 8.2.

8.2.3 INTERPRETATION OF THE RESULTS

There is no way in which the results can be made to refer to a single chemical composition for the entire Earth because the pressures involved are never great enough to force the required degree of compression. Rather, the mantle–crust region can be regarded as broadly related chemically while the core must be different and heavier. Reducing the calculated magnitudes for the densities to those corresponding to zero pressure, and reviewing the results in the light of the known cosmic abundance of the elements it becomes clear that the mantle–crust must be composed primarily of silicates while the core region must be primarily of ferrous composition, as assumed before. This would mean that ferrous core materials form about 32% by weight of the total material: explicitly we find:

$$\text{Mass (core)} = 1.9 \times 10^{24}\ \text{kg} \qquad (32\%)$$

$$\text{Mass (mantle)} = 4.1 \times 10^{24}\ \text{kg} \qquad (68\%)$$

$$\text{Mass (crust)} = 2.4 \times 10^{22}\ \text{kg} \qquad (0.4\%)$$

We see that the mass of the core is four times that to be expected on the basis of a cosmic abundance alone (see §8.1). The terrestrial crust accounts for only about one per cent of the terrestrial mass (coincidentally, almost exactly the same magnitude as the Moon) and so is too small a region to have any effect on the equilibrium of the interior itself.

The density will generally increase with depth, corresponding to a changing silicate and ferrous composition, and the seismic data show a fine structure of layers within the Earth. These will be determined primarily by chemical affiliations, influenced both by the present temperature distribution within the Earth and by the initial composition and early history. The seismic wave speeds allow the rigidity and bulk modulus of the material to be deduced as a function of the depth and this information can be used as a guide to the material composition likely to be there.

Conditions in the interior of the Earth are controlled primarily by the degree of compression (see chapter 2), so seismic measurements are not critically affected by the temperature and cannot be used to distinguish between the isothermal model considered so far and thermal models that would perhaps be more realistic in fine detail. The time scale of seismic disturbance (seconds or minutes) does not match that of expected internal convection (10^8 years) so here again seismic disturbances will not couple with the thermal regime through which they pass.

The Earth's magnetic field is to be associated with a liquid, electrically conducting region (see chapter 6) and the outer core meets these criteria precisely. If the source of energy for the field is convection induced by solidification we would infer that the extent of the inner core has increased with time, quickly at first but now more slowly. Certainly the particular terrestrial core structure cannot necessarily be expected as a general feature of a terrestrial-type planet. It is interesting that the division between the solid mantle and liquid outer core coincides with a gross difference of material composition, but this would seem not necessarily to be the case in general. Indeed, it is not the case for the division between inner and outer core. The solidification of the central region of the Earth must be the result of the particular dependence of the melting point of the ferrous materials on pressure, although it could be an effect involving the gross increase of viscous dissipation.

The structure indicated so far (see figure 8.3) is consistent with the deductions from observations of the free oscillations of the Earth. This is a more recent development that offers definite prospects for the future.

8.3 GENERAL PLANETARY MODELS

Instead of beginning with observational data, we can construct a series of models on the basis of a range of compositions to see the general properties of

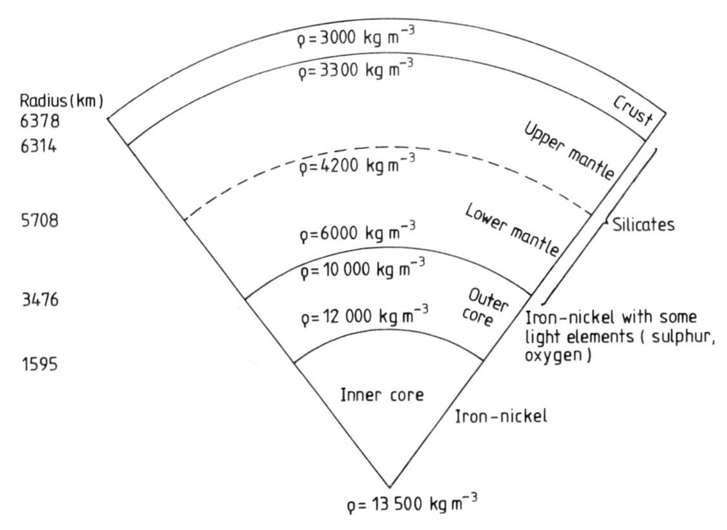

Figure 8.3 *The internal structure of the Earth.*

planetary bodies of terrestrial type from the point of view of the requirements of hydrostatic equilibrium. This approach will presuppose an equation of state of the material, and such equations were considered in chapter 3. The simplest, and probably most generally useful, equation of state is the Murnaghan equation (3.20) or (3.21) and we use this now. The equation for the density is the Emden equation (3.33) which is now to be solved subject to the boundary conditions that the total mass and mean density are known, and that the pressure at the surface is zero. The method presupposes that the gross internal structure is assigned; the distinction between a liquid or solid core is to be made on the basis of corresponding small differences of density between the two cases. Alternatively, the planetary mass can be used as the controlling variable of the calculation by appealing to equation (3.27) and the boundary conditions (3.30).

8.3.1 NEARLY HOMOGENEOUS PLANET

An elementary approach to the study of the interior is possible when the degree of compression is small and the chemical structure is close to homogeneity. Then the inertia factor is only slightly smaller than 0.4, applying to a homogeneous body, and the formulae can be applied in a simple way. Such an analysis can be applied, for instance, to the Moon, where the inertia factor is about 0.392 (see figure 8.4). The most direct approach is in terms of the equation for the mass (3.27). This approach has been applied to preliminary studies of the Moon and to Mars. The mass at distance r is expressed in powers of the parameter θ and the separate coefficients determined by solving the

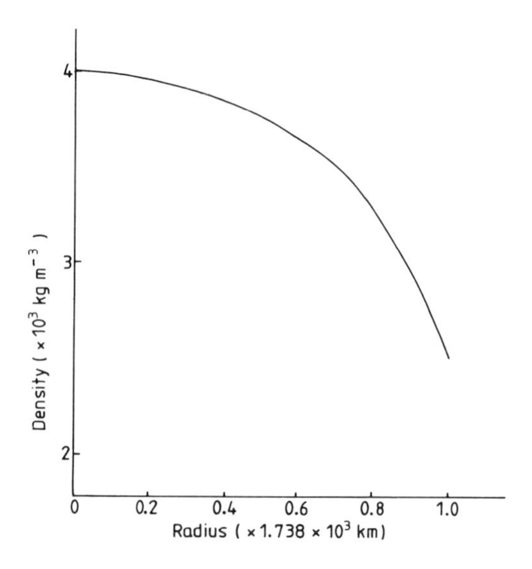

Figure 8.4 *Dependence of density on depth within the Moon, supposed to be composed of Murnaghan material with* B = 4. *Mean density* = 3342 kg m^{-3}, *inertia factor* = 0.392. *The inertia factor would be reduced if a ferrous core were added to the model.*

appropriate set of differential equations analytically. Alternatively, the equation can be solved numerically given the bulk modulus as a function of the depth. In this case, the equation for the mass is best solved by specifying both m and dm/dr at the centre, dm/dr being proportional to the central density. The solution is carried through until the pressure for a given r vanishes. This is the condition for the surface, and the appropriate r is then the radius of the planet R_P. The magnitude of the radius found this way will depend on the central density, as will the enclosed mass. The calculation is continued for a range of magnitudes for the central density and that particular magnitude is chosen which provides the observed magnitudes for both mass and radius (and so for the mean density of the body). With the distribution of the density with depth known the inertia factor can be calculated. In this way, the internal conditions for the planet are determined (see figure 8.5).

There is not a unique internal structure for a given set of parameters (R_P, M_P, α_P). Whereas the silicate–ferrous composition expected of a terrestrial-type planet will prescribe a set of models within certain limits, the precise composition will remain vague to some extent. It will be possible, however, to specify a range of materials through the expression used for the bulk modulus that will be consistent with a given pair of mass–radius values. This will allow the interior structure to be specified in broadly statistical terms rather than in terms of particular composition and the approach has been explored by Press.

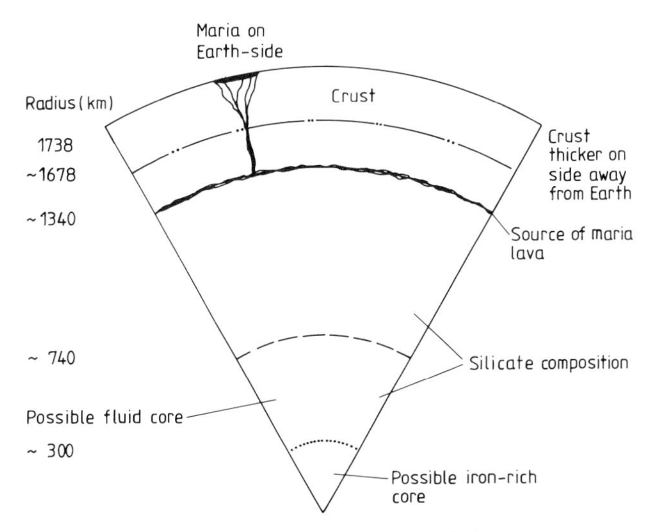

Figure 8.5 *The internal structure of the Moon.*

One unknown in all these calculations is the condition of the ferrous material inside. Whether it is in combination with the silicate elements in some proportion or whether the bulk of it is free is something that cannot be determined uniquely from outside. The thermal conditions will be relevant here but these are not known with any precision. There is a substantial element of uncertainty associated with the solution of the differential equation for the mass (or equivalently for the density), although it is clear that the observed data are not consistent with cores as small as would be indicated from the cosmic abundance alone (see §8.1). Assuming all the ferrous components are collected in a central core, the models indicate the following proportions of the masses to be allocated to the cores of the different planets and to the Moon and Io:

Planetary body	Percentage of radius	Percentage of mass
Mercury	70	90
Venus	45	29
Earth	50	32
Mars	15	20
Moon	<31	<7.5
Io	<45	<18

These data (with the exception of those for the Earth) have no *in situ* observational support and seismic data for other planets are necessary to quantify these deductions.

8.3.2 CONDITIONS IN THE UPPER MANTLE AND CRUST

The surface conditions are particularly affected by the conditions in the upper mantle and crust and an estimate of the level of activity in the surface region can be made on the basis of mountain height.

To see the principle, suppose the surface is plane and that the mountain (made of the same material) is a rectangular object of height H and with a base of unit area. The mountain is supported by the surface and the higher the mountain the greater the pressure at the base to be supported by the surface. The strength of the surface is determined by the binding energy of the constituent atoms; the greater the binding energy the greater the strength and the higher the mountain that can be supported. As more material is added to the mountain the height will increase but there will be a condition (which determines the maximum height) when the addition of even one more atom will exceed the binding strength of the base. The base will then show plastic flow. At this stage the mountain will sink by an amount to eliminate the excess energy which breaks the interatomic bond. To an approximation sufficient for our present purposes we can suppose that when the mountain sinks the volume of material that sinks into the base is the same as that placed on the top causing the sinking. It is, on this basis, possible to relate each atom added to the top of the mountain with a corresponding atom that sinks into the base as a result. This relates the maximum height of the mountain to the interatomic binding energy of the material.

The potential energy of an atom of mass Am_{p} at height H in the gravitational field with acceleration g is $Am_{\mathrm{p}}Hg$ and this energy is to be equated to the binding energy of the atom E_{b}. This gives the condition for maximum height

$$Am_{\mathrm{p}}Hg = E_{\mathrm{b}}.$$

It is interesting to notice that Hg=constant for a given material; since g decreases as the mass decreases, smaller planetary bodies are associated with greater maximum mountain heights. Indeed, the minimum size for a spherical planetary body can be taken as that magnitude of the radius which is of the same order as the maximum height of a possible 'mountain'; this gives the minimum radius of about 100 km, in agreement with the estimate of chapter 2.

For most silicate materials, the binding energy per atom has a magnitude of order 10^{-3} Ryd, where Ryd is the Rydberg energy introduced in chapter 2. The maximum mountain height is then given by the condition

$$Am_{\mathrm{p}}Hg \sim 10^{-3} \tag{8.2}$$

from which H can be deduced when the other variables are given.

The formula for the maximum height can be expressed in an alternative form involving the size of the planet. The surface acceleration of gravity itself depends on the amount of matter in the planet, and so on the type of material, the amount of it and the physical condition (expressed through its size).

Explicitly, for a spherical body of radius R_P and mass M_P, we write

$$g = GM_P/R_P^2 \tag{8.3}$$

where G is the constant of gravitation. The mass can also be written in terms of the proton mass m_p. It is a coincidence that the atomic masses of silicates and iron are quite similar, so to a sufficient first approximation we need not distinguish between them. We will set $A=50$ as an acceptable approximation for our present purposes. If there are N_P nucleons within the planet as a whole

$$M_P = m_p N_P. \tag{8.4}$$

Supposing each atom to have a radius $r_a a_0$, where a_0 is the Bohr radius, we can write to the present approximation

$$R_P = r_a a_0 (N_P/A)^{1/3} \tag{8.5}$$

Combining equations (8.2), (8.3), (8.4) and (8.5) we obtain the expression for the maximum height H

$$H/a_0 = a_0 10^{-3}\ \mathrm{Ryd}(Gm_p^2 A^{5/3} N_P^{1/3})^{-1}$$

in terms of the Bohr radius. We remember the expression (2.11b) for Ryd and so obtain the final expression

$$H/a_0 = 10^{-3}(a_1/a_2)N_P^{-1/3}A^{-5/3} \tag{8.6}$$

where

$$a_1 = e^2/hc \quad \text{(the atomic fine-structure constant)} \tag{8.6a}$$

$$a_2 = GM_P^2/hc \quad \text{(the gravitational fine-structure constant).} \tag{8.6b}$$

The ratio a_1/a_2 is the ratio of the electric to gravitational energies for two protons. This ratio expresses the maximum height as the balance between gravitational and electric forces. We can notice that the same balance is present in other everyday things like raindrops (on Earth!) and ripples on a lake due to the wind. Because the balancing forces are the same in these different cases, a relation exists between their relative magnitudes (it follows that the maximum size of a water drop is related to the maximum height of the mountain!).

8.3.3 APPLICATION TO TERRESTRIAL PLANETS

Calculated data for the terrestrial planets are collected in table 8.1. Also included is the maximum height of a mountain on the largest possible (hypothetical) iron planet and also for a neutron star (composed of hydrogen).

Table 8.1 *The calculated maximum height of a mountain on each of the terrestrial planets compared with the heights actually observed. Data for a hypothetical largest planet of iron composition and for a neutron star are included for interest.*

Planet	Surface gravity ($m\,s^{-2}$)	Calculated maximum height H (km)	Observed maximum height H_0 (km)	Ratio H/H_0
Mercury	3.98	63.9	—	—
Venus	8.60	29.6	11	0.37
Earth	9.78	26	8	0.31
Moon	1.62	157	6	0.038
Mars	3.72	68.3	26	0.38
Largest Fe body	2.79×10^3	10^{-1}	—	—
Neutron star	10^8	2×10^{-6}	—	—

The latter data show that a neutron star is extremely smooth and that the roughness on the largest possible terrestrial-type planet would be difficult to see at any distance. For the planets actually observed it is extraordinary that the ratios of observed to calculated values turn out to be essentially the same for Earth, Venus and Mars. The comparison of the heights themselves would be very untrustworthy as the basis for informed speculation but the comparison of the ratios of observed to calculated heights would seem safer. This comparison suggests that the level of crustal activity in the three planets was the same when the observed features were formed. This does not mean that the detailed mechanisms are the same, and we cannot infer that plate tectonics apply to Venus and Mars as well as the Earth. The particular manifestation of the surface thermal boundary layer (see chapter 6) will depend on particular circumstances (for instance, the amount of water in the minerals) and the occurrence of plates is to a large extent coincidental. Indeed, the Earth has not shown this form of tectonic activity throughout its history. The fact that the maximum terrestrial elevation is due to lifting due to plate collisions whereas the high regions of Venus and Mars appear due to volcanic activity can certainly be noted. The level of activity on the Moon is very low, suggesting the lunar crust has never been subject to substantial tectonic activity. The heights of features on Mercury are only incompletely known, even over that portion of the surface that has yet been observed. The general surface appearance is similar to that of the Moon, and if that is so the theory we have discussed would suggest a maximum height of some 2.5 km for the elevations on Mercury. Later space missions will allow the full surface of the planet to be surveyed and the maximum height determined.

8.4 MAGNETIC DATA

The magnetic conditions of the terrestrial planets were referred to in chapter 6. Only the Earth and Mercury are known to have general fields of internal origin. Whereas the terrestrial field has been studied in great detail over an extended time period, that of Mercury has been sampled only as part of the magnetic environment by a fly-by spacecraft. It seems clear, however, that the strength of the Earth's field is some 100 times that of Mercury. This difference is interesting and it could be that the particular liquid/solid conducting core structure found in the Earth is conducive to a strong field production, while the conditions in Mercury (where this structure is not expected) are less favourable from the magnetic point of view.

The magnetic environments of Mars, Venus and the Moon are more enigmatic. The strength of any general Martian field is apparently less than one thousandth that of the Earth, while for Venus it must be less than one ten thousandth and for the Moon less than one hundred thousandth. The Moon shows palaeomagnetism (see chapter 6).

At least some of the satellites of the major planets may have intrinsic magnetic environments. There is some evidence that Io has an internal field although the field perturbation observed by the Voyager fly-by was caused by the general Jovian environment. Nevertheless, it is likely that an upper limit can be placed on a magnetic field of internal origin of about one tenth of the terrestrial field strength. For Titan, an upper limit of about one thousandth of the Earth's field strength might be an appropriate magnitude. The magnetic conditions for the other satellites have not yet been explored in sufficient detail to allow meaningful conclusions to be drawn. It can be concluded quite generally, however, that the recognition of a magnetic field of internal origin is sufficient evidence for differentiation of the constituent material into an electrically conducting iron core and an enclosing silicate mantle. The precise conditions in the core remain unknown.

It remains for the future to interpret these and other observations in detail in the light of the arguments developed in chapter 6.

8.5 CONCLUSIONS

1 The terrestrial planets are composed of the least abundant elements in the cosmic abundance list.

2 The spread of densities and general properties suggests that they are made of silicates and ferrous materials, the proportion of the two components being different from one member to another.

3 The interior structure of the Earth has been investigated in detail using seismic waves as the probe. The inner half (the core) of the volume is found

to be a high-density material and iron is expected to be a major component. This is encased in a silicate mantle and the whole is covered by a thin crust of light silicates.

4 This general structure is likely to be representative of all the members of this group (the Moon possibly also has a small iron core). The proportions of material in the core and the mantle will differ from one body to another.

5 The core of the Earth has a particular structure. There is an outer half which cannot withstand shear (liquid) and the inner half has a high rigidity, comparable to that of a solid. This liquid/solid structure is probably unique.

6 Conditions in the upper mantle cannot be inferred uniquely from outside but surface features can offer clues. One clue is the magnitude of the heat flux. Another is the observed height of surface features, to be compared with theoretical estimates of the strength of the upper mantle.

7 The magnetic environment of the Earth has been studied in ever more detail over the last three hundred years. The general main magnetic field of internal origin has its origin in the liquid region of the outer core. The core structure for the Earth is probably unique. The terrestrial field is also the strongest observed for this type of body.

8 Mercury is known to possess a magnetic field but the magnetic environments of Venus and Mars are still obscure. If magnetic fields exist they have strengths less than 10^{-3} of the terrestrial strength. The Moon has evidence of a palaeofield but not a field now (at least not as strong as 10^{-5} of the terrestrial field strength).

9 Io and Titan show evidence of having internally maintained magnetic fields. The strengths are unknown except for being less than 10% of the terrestrial field for Io and less than 0.1% for Titan.

10 These magnetic data promise to be of great future importance to the dynamo theory of the field production.

REFERENCES AND COMMENTS

The literature about the terrestrial planets is dominated by that for the Earth. It is not possible to give a comprehensive literature survey but we list here various books and papers, and add a variety of comments, to allow the reader to delve further if desired.

The classic book for the Earth is

Jeffreys Sir Harold 1976 *The Earth* 6th end (London: Cambridge University Press)

This book is a mine of information and is very well and clearly written.

A book giving extended details of the interior structure with the methods by which it has been obtained is

Bullen K E 1975 *The Earth's Density* (London: Chapman and Hall)

An excellent general introductory text is

Garland G D 1979 *Introduction to Geophysics* 2nd end (Philadelphia: Saunders)

A general and informative text is

Brown G C and Mussett A E 1981 *The Inaccessible Earth* (London: Allen and Unwin)

A general geologically orientated book is

Bott M H P 1982 *The Interior of the Earth: its structure, constitution and evolution* 2nd edn (London: Edward Arnold)

A survey of matter relating to the Earth's core is

Jacobs J A 1975 *The Earth's Core* (New York: Academic)

A reference for the Moon is

Taylor S R 1975 *Lunar Science: A Post Apollo View* (Oxford: Pergamon)

Where no specific reference to a detail is made these books may be consulted for it.

The historical beginnings of the subject are not without interest. Among the first to speculate on the interior structure of the Earth was Wiechert in

Wiechert E 1897 *Nachr. Ges. Wiss.* Gottingen p 221

He realised that the high mean density of the Earth could not be understood on the basis of the simple compression of surface rocks of only roughly half the mean density. He further observed that the meteorites (the only extraterrestrial material known to him) could be put into two compositional classes, one being rock with a density not dissimilar to terrestrial surface rocks, and the other iron with a density roughly twice as great. The non-homogeneous model suggested itself of a heavy iron core of outside radius $R_c = xR_P$ (say) with density 8200 kg m^{-3} encased in a rock mantle. Wiechert supposed the density of each region to be uniform and found the data gave $x = 0.8$ for the Earth. Seismic studies gave instead $x = 0.545$ and Jeffreys pointed out that Wiechert went wrong when he did not account for the compressibility of the material. However, the model has been useful in the evolution of our ideas about the interior structure and can still be useful as a preliminary guide, particularly when the inertia factor is known.

Suppose ρ_m, ρ_c and ρ_P are the densities respectively of the mantle, core and entire planet. Suppose further that the inertia factor α_P for the planet can be assigned, for instance by combining data for J_2 with measurements of the precession of the equinoxes. Considerations of mass and moment of inertia then lead, respectively, to the two relations

$$\rho_m + x^3(\rho_c - \rho_m) = \rho_P$$

$$\rho_m + x^5(\rho_c - \rho_m) = \tfrac{5}{2}\alpha_P \rho_P$$

from which ρ_m and ρ_c can be determined when α_P and x are assigned. For the Earth, $x = 0.545$ and $\alpha_P = 0.331$, giving $\rho_m = 4220$ kg m^{-3} and $\rho_c = 1233$ kg m^{-3}.

The assumption that the core is made of iron has been modified over the years to include other impurities for empirical reasons but the overwhelming ferrous nature of material is still generally accepted. The possibility that the core is made instead of a high density modification of silicate material, with at most a small ferrous impurity, has been considered especially by Ramsey:

Ramsey H W 1948 *Mon. Not. R. Astron. Soc.* **108** 406

While the ferrous composition is now accepted it must be admitted that the final argument to discredit a silicate core has not yet been found. Perhaps it is more difficult to demonstrate that something does not exist than that something does.

RC8.1 GENERAL CONSIDERATIONS

The relation of the composition to the cosmic abundance is still rather speculative, but is treated in a growing number of books and papers. One of interest is

Hoyle Sir Fred 1978 *The Cosmogony of the Solar System* (Cardiff: University College Cardiff Press)

RC8.2 APPLICATION TO THE EARTH

RC8.2.1 Seismic waves. Seismic studies of earthquake disturbances are based on the assumption that the material of the Earth behaves closely like a perfectly elastic solid over periods of minutes or hours. The length of the waves generally studied is of the order of a few kilometres with associated periods of a few seconds. This corresponds to wave speeds in the general range 3–10 km s^{-1}.

Two general modes are recognised for the propagation of elastic energy within an infinite medium. One is longitudinal (a compressional sound wave, usually called a P-wave) and the other is pure transverse (a shear or S-wave). The P-wave speed V_P is determined by the density ρ, the bulk modulus K and rigidity modulus μ of the medium, while the S-wave speed V_S is determined by the density and the rigidity. Explicitly

$$\rho V_P = K + \tfrac{4}{3}\mu$$
$$\rho V_S = \mu. \tag{RC8.1}$$

It follows that $V_P > V_S$.

Whereas both waves propagate if the rigidity is not zero, only the P-wave propagates if it is. This means that a region where both waves' modes propagate is solid (or at least has very high viscosity) while that where only the P-wave propagates is liquid (unable to withstand shear). This distinction has been of the greatest importance in geophysics. For a self-gravitating sphere the same results apply, except that now gravity affects the trajectories of the P-waves, curling them to a degree dependent on the strength of the field; the S-waves will not be affected. These matters were studied early on by

Love A E H 1911 *Some Problems of Geodynamics* (New York: Dover)

Inhomogeneities will also distort the wave pattern. A surface of discontinuity between two different elastic media will cause an incident wave to be refracted and reflected according to the rules familiar in optics, but now both secondary waves will show both P and S characteristics, irrespective of whether the primary (incident) wave was pure or mixed. This adds a great element of complication to the analysis of wave trajectories.

Conditions near a free surface are different again. Within about ten wavelengths of the surface (i.e. within a depth of some ten to twenty kilometres), a pure longitudinal component does not propagate. Instead, a mixed longitudinal and transverse wave with vertical polarisation is found, called a Rayleigh wave. The amplitude decreases exponentially with the depth and the wave is analogous to the surface wave of hydrodynamics. If the material is suitably layered, a pure transverse wave is possible with horizontal polarisation (i.e. parallel to the surface). This is the Love wave. Other propagation modes are possible in the surface region under particular circumstances.

The elastic volume will show collective oscillations under the action of an impulse (for instance an earthquake), as a drum vibrates when beaten or as a bell rings when struck by a hammer. For planetary studies it is the behaviour of a spherical volume that is especially important. These collective modes were first investigated theoretically by Love (see the reference above) in 1911 but were recognised in practice for the Earth only thirty years ago through the study of strong earthquake shocks. Both longitudinal and shear body oscillations occur, and the frequencies of the oscillations show a discrete spread depending on the internal structure. Frequency splitting arises from rotation and has been identified for the Earth. The body oscillations involve the total volume and contain the full range of possible vibration modes. Interference effects between the different modes give rise to localised resultant disturbances and the P and S disturbances can be regarded as interference waves from general modes of high order (typically in excess of 100).

The theory of body oscillations can be studied in the book by Bullen quoted above and also in

Lapwood E R and Usami T 1981 *Free Oscillations of the Earth* (London: Cambridge University Press)

RC8.2.2 Density and pressure. There is a vast literature on this subject. A good survey is contained in the book by Bullen quoted above, much of this work having been done by Bullen and his research students.

RC8.3 GENERAL PLANETARY MODELS

Most of the models are based on isothermal equations of state. The methods can be used in principle to find some information about the elastic properties of the material; see for example

Cole G H A 1971 On Inferring Elastic Properties of the Deep Lunar Interior *Planet. Space Sci.* **19** 929–47
Cole G H A and Parkinson D 1972 On Inferring the Elastic Properties of the Deep Planetary Interior: Moon and Mars *Planet. Space Sci.* **220** 557–69

More realistic thermal conditions can be included. One way of achieving this is by way of the thermal pressure introduced by Anderson:

Anderson D L 1967 *Geophys. J. R. Astron. Soc.* **13** 9
—— 1979 *Phys. Chem. Miner.* **5** 33
Anderson D L and Sumino Y 1980 *Physics of the Earth and Planetary Interiors*

The idea is to separate the pressure into a part which applies at zero temperature, $P_1(\rho)$, and which depends on the density and elastic properties (bulk modulus) but not the temperature, and a part $P_2(T)$ which depends explicitly on the temperature. For the total pressure we write

$$P(\rho, T) = P_1 + P_2 \qquad \text{(RC8.2)}$$

and P_1 can be assigned an isothermal form such as is considered in chapter 3. Although this part strictly applies at the absolute zero of temperature the planetary interior is presumed cold and the difference of parameters between zero temperature and that actually occurring will not be significant.

The form of the thermal part is determined from thermodynamics. We write

$$P_2(T) = \int \frac{\partial P}{\partial T} \, \mathrm{d}T$$

$$= \int K\beta \, \mathrm{d}T$$

where K is the bulk modulus and β the coefficient of thermal expansion for the material. Above the Debye temperature it is found empirically that $K\beta$ is virtually independent of the temperature. It follows that

$$P_2(T)=aT-b$$

where a and b are constants of integration.

The constant a is determined by the bulk modulus and thermal expansion coefficient according to

$$a=K(\rho)\beta(\rho).$$

The magnitude for b is obtainable from laboratory data.

This approach has been used to investigate the internal thermal conditions in terrestrial planets by

Baumgardner J R and Anderson O L 1981 Using the Thermal Pressure to Compute the Physical Properties of Terrestrial Planets *Adv. Space Res.* **1** 159–76

The isothermal equation of state is of the Born–Mie form and the components of the calculation were

(i) equation of state for the material
(ii) equation for hydrostatic equilibrium
(iii) moment of inertia of the body
(iv) gravitational acceleration
(v) heat flow in a spherical body

$$\mathrm{d}Q/\mathrm{d}r=H-2Q(r)/r$$

(vi) distribution of heat generation in surface region

$$H_s=H_0 \exp[-(R_P-r)/D]$$

(vii) thermal gradient within the body

$$\mathrm{d}T/\mathrm{d}r=-Q/K(r)$$

(viii) temperature distribution in convecting shells

$$\mathrm{d}T/T=\gamma\,\mathrm{d}\rho/\rho$$

where γ is the Grüneisen parameter

(ix) the variation of magnitude of the Grüneisen coefficient with density

$$\gamma/\gamma_0=(\rho/\rho_0)^q$$

where q is a constant parameter

(x) the superadiabatic temperature difference across a virtually adiabatic shell

$$\beta(\Delta T)_{\mathrm{supad}}=\int_{\rho_1}^{\rho_2}(\mathrm{d}\rho/\rho-\mathrm{d}p/K_S)\,\mathrm{d}\rho$$

(xi) the solidus curve for the mantle material

$$T_m K = 1500 + 10P \qquad (P \text{ in GN m}^{-2})$$

The results of calculations based on these formulae provide data in reasonable agreement with those derived from seismic data and temperature distributions consistent with surface measurements. The central temperatures are a few times 10^3 K.

The modelling does, however, need to make assumptions about the internal distribution of free ferrous materials. Seismic measurements for the terrestrial planets are a desperate present need for future developments of the theory. The structural parameters are, of course, hypothetical at the present time. Calculations of this type are important in providing a background 'feel' for the problem, which will prove invaluable when more reliable data eventually become available.

For comparison, isothermal calculated data are also included in the paper of Baumgardner and Anderson for Earth, Moon and Mars.

9

PLANETARY BODIES COMPOSED MAINLY OF ICE

Hydrogen and oxygen are the first and third most abundant elements in the Universe. They also combine together readily to form water. Carbon and nitrogen are respectively the fourth and fifth most abundant elements, and these too combine with hydrogen particularly to form methane (CH_4) and ammonia (NH_4). It might be expected, therefore, that water, methane and ammonia would be very obvious constituents of the Solar System, but viewed from the Earth this does not seem to be the case. It is true that the Earth is covered with the oceans but the amount of water in these is not as great as would be expected on the scale of cosmic abundance. The water in the terrestrial atmosphere accounts for only about 0.01% of the total by number of atoms. Water is very noticeably absent on the Moon and it seems so also on Mercury. It is a minor constituent in the atmosphere of Venus (about 100 parts per million) and of the Martian atmosphere (about 300 parts per million). There may well be frozen water in the crust below the Martian surface. These various sources of water in the inner Solar System involve only relatively small quantities and do not account for the amounts that might be expected on elementary arguments. Instead, it seems that hydrogen and oxygen have combined there in more complicated circumstances, particularly involving carbon, to form carbonates of elements. Methane and ammonia are known to be present in the Jovian (and presumably Saturnian) atmospheres, but again the quantities are small and not of a magnitude compatible with the raw solar abundance.

The results of certain spectroscopic analyses of the satellites of Jupiter and Saturn made long ago using Earth-based telescopes led to early speculations that water-ice is a major surface constituent of these bodies, but the recognition of the true situation came only with the recent data sent back by the Pioneer fly-bys of the major planets. The Galilean satellites of Jupiter (excepting Io) and the larger satellites of Saturn have substantial water-ice mantles. The less dense and smaller satellites will be composed almost entirely of ice. More than this, as we saw in chapter 7, Uranus and Neptune could well have water as a major internal constituent themselves. Pluto also is likely to have a substantial ice mantle.

The reason why water is able to play so dominant a role but methane and ammonia cannot is associated with the melting and boiling temperatures of these various compounds. The melting temperature of methane is 89 K (boiling point 111.7 K) while that for ammonia is 195.8 K (boiling point 239.8 K). The internal temperatures of the satellites are likely to exceed 300 K from which it follows that methane and ammonia ices are unlikely to be dominant stable constituents of the satellite bodies except in restricted regions of the outer mantle and crust. Only water-ice is stable over the range of temperatures likely to be met (see §9.4) throughout the volume of the mantle. Apparently, the regions of the Solar System beyond some 10 AU from the Sun are very rich in water-ice; indeed, ice is to be regarded as the main rock-forming mineral of the solid planetary bodies in this region. The geology of the satellites of the major planets is then very much the study of the behaviour of ice under various conditions of temperature and pressure. It is important, therefore, from a theoretical point of view, to investigate the broad physical properties of a planet composed primarily of water-ice. For this, the rheological properties of ice must be known in detail at the low temperatures (<150 K).

9.1 PHYSICAL PROPERTIES OF WATER-ICE

The ice with which we are familiar on Earth (ice I_h) is only one of ten polymorphic forms. It is the form stable at room temperature and atmospheric pressure and its properties are known with some precision. It has a hexagonal atomic structure with a Wurtzite-type lattice for the oxygen ions. It is a very brittle material at temperatures below 200 K, even when confined by a pressure of 100 M N m^{-2}. At such a high pressure it becomes plastic only for a strain rate in excess of 4×10^{-6} s^{-1}. For high dynamic stresses it fractures by cleavage, showing no ductile behaviour.

It is convenient to normalise the shear stress to the shear modulus and the temperature T to the melting temperature T_m. For $T = 0.9 T_m$ and a shear of 10^{-4} μ, the creep rate is of the order 10^{-10} s^{-1}. This is smaller by a factor 10^3 than the rate for the covalent elements silicon and germanium, and a factor 10^6 smaller than for the alkali halides. The magnitude of the shear modulus is not known but (by analogy with the alkali halides) it may be relatively low in comparison with other materials. The kinematic viscosity ν is not known accurately over a wide range of conditions but several authors have used the dependence (in SI units) on the temperature

$$\nu = 1.39 \times 10^{-5} \exp (7.214 \times 10^3/T).$$

This is an application of the usual empirical formula exp (E/kT) (where E is an activation energy, k the Boltzmann constant and T the temperature) expressing the dependence of liquid viscosity on temperature.

Ice I_h is one of the strongest materials known and is a good material for rock construction. It is unlikely to be found in pure form in the satellites and the effects of impurities require investigation. While it might be expected on crystallographic grounds that impurities will tend to soften the ice, it is also known empirically that the effect of ammonia in solution is to harden it, although the reason is by no means clear. An important impurity will be silicate materials and this can be simulated generally by considering ice–sand mixtures. Results of an investigation reported by Lange and Ahrens are given below for strain rates of the order $10^3\ s^{-1}$. The presence of sand apparently increases the strength substantially, but the effect of temperature (both in allowing differentiation, which reduces the impurity level, and in allowing flow itself) can be an important and significant variable in planetary applications. One effect of the silicate impurity in an undifferentiated body will be to inhibit dynamic recrystallisation and grain growth, effectively softening the ice. Another effect is to cause dispersion hardening, the creep rate decreasing but the activation energy increasing with increasing volume fraction of sand. Estimates of the proportion by mass of rock material likely to be found in the satellites are given in RC9.3.

Table 9.1 *Tensile strength of pure ice and ice–sand mixtures.*

Material	Tensile strength (MN m^{-2})
Pure Ice	16–17
Ice–5% sand	18–20
Ice–30% sand	22–25

At temperatures below about 200 K, I_h (which has the hexagonal stacking structure ABAB) is metastable at low pressures ($< 2\ GN\ m^{-2}$), the stable ice form being cubic ice I_c. Very little is known of its mechanical properties but the similarity between I_c and I_h, on the one hand, and between I_h and Si and Ge (Ge also expands on solidification as does ice) on the other suggests similar mechanical properties between I_c and Si and Ge. This would make the ice a very strong solid. Ice can behave as a proton semiconductor under suitable conditions.

Various polymorphs of ice arise at high pressures but the associated mechanical properties are largely unknown in detail. What information there is involves inferences drawn from crystallographic expectations. What is at present known of the phase diagram is given in figure 9.1.

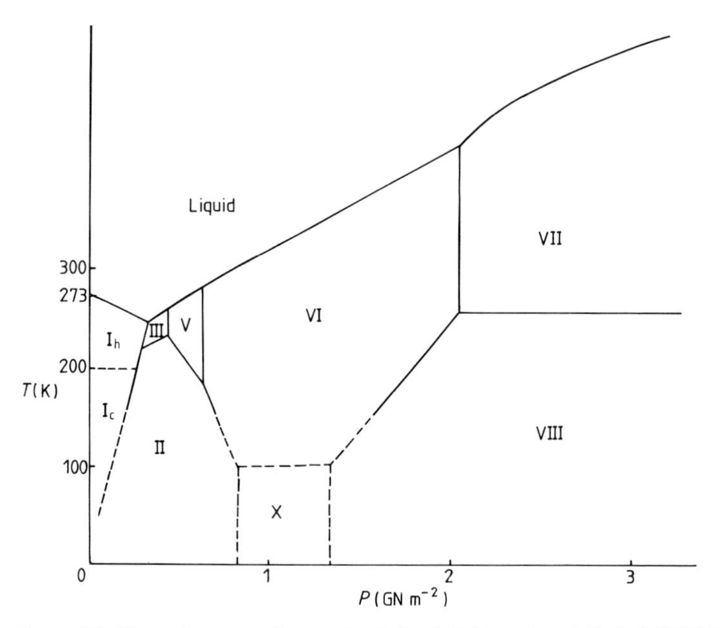

Figure 9.1 *Phase diagram of water-ice (after Mishima O and Endo S 1980* J. Chem. Phys. **73**).

Ice II has a rhombohedral form and is more densely packed than ice I. While the density ρ_1 of ice I is about 920 kg m^{-3}, that of ice II is some 25% higher: $\rho_{II} \sim 1.17 \times 10^3$ kg m^{-3} at 10^5 N m^{-2} and 100 K. There are indications that the viscosity of ice II is greater than that of ice I (where the regions of stability overlap) although the quantitative difference is still unclear. Ice II has no melting point, but transforms to ice III or ice V (see figure 9.1). At the same temperature, ice III is softer than I_h by a factor of about 10^3.

Ice VI is stable at room temperatures in the pressure range 0.55–2 GN m^{-2}, although the upper limit falls as the temperature goes down. It has the structure of two identical and independent interpenetrating chains of tetrahedral structure linked by water molecules to give a tetragonal structure in the c-direction. For temperatures close to the melting point and shear stresses of order 10^2 bar the viscosity is of order 10^{13} N s m^{-2} according to laboratory measurements. The shear modulus is unknown but its closer packing than ice I suggests a higher modulus than for I.

The densest of all the polymorphs is ice VII with a density about 1.5×10^3 kg m^{-3} at a pressure 10^5 N m^{-2} and temperature $T = 100$ K. It has a cubic crystal structure comprising two interwoven lattices of I_h with distorted protons. The protons order at low temperatures to form ice VIII, the dipoles being directed antiparallel in each lattice to give an antiferroelectric ordering.

This summarises the principal general information available for ice at the present time and more information is urgently needed. It could be that the study of the icy satellites themselves will allow more of the properties of ice to be inferred.

9.2 GENERAL CONSIDERATIONS

The characteristic features of an icy satellite must be a low density (around 1000 kg m^{-3} for virtually complete ice composition, but certainly less than 3000 kg m^{-3} for an ice–silicate mixture) and small size (generally smaller than Mercury although the Galilean satellites are comparable in size or larger, as are Titan and Triton). The observational data for the satellites of the major planets are contained in tables 1.6 and 1.7 and the data for Pluto in table 1.1.

The surface features will be different for ice than for silicate compositions. For instance, a crater in a silicate surface will be convex downwards as a general rule whereas the opposite is observed for ice. For ice the surface features will be determined by expansion on freezing (for ice I) rather than the contraction more familiar for the terrestrial bodies. Proud extrusions of ice will be seen and the elucidation of the history from surface features will be more complicated.

An old surface will be highly reflecting with very low relief. A series of lineaments will be associated with regions of particular and extended strain and low hills will be associated with extrusion from below. Craters will relax in a way dictated by the viscosity of the ice. This will depend on temperature and on the impurity level of the ice. Some silicate impurity will strengthen the ice and smaller craters will result from impact as a result. Here the surface details give a clue to conditions inside. For instance, on the small body Mimas the large crater has retained its shape and sharpness whereas the large crater on the larger body Tethys has already been reduced to a circular scar. The implication is that the interior of Tethys is warmer than that of Mimas. The Valhalla crater on Callisto and the marks of craters on Ganymede are extreme cases of the results of ice deformation. Large craters are often surrounded by a ray structure or a ring structure which are presumably the result of the brittle response of crustal ice to tension. An ability to read the surface features more competently is a pressing need at present.

9.3 MODEL SATELLITES

There is considerable uncertainty in detail about the internal physical structure of the bodies and this must be reflected in the general form of current models.

There are three choices; one is that the body is totally differentiated with a rock core and an ice mantle, and the data for this are collected in tables RC9.3 and RC9.4 of RC9.3. A second choice is that the material is fully mixed. The results of such an analysis are also contained in tables RC9.3 and RC9.4. The third choice is a mixture of the two, the material showing only partial differentiation. The same structural possibilities apply to Pluto, the outermost planet, but the arguments of RC9.4 suggest the planet is most likely differentiated: a possible model is shown in figure 9.2 as a representative example of an icy planetary body.

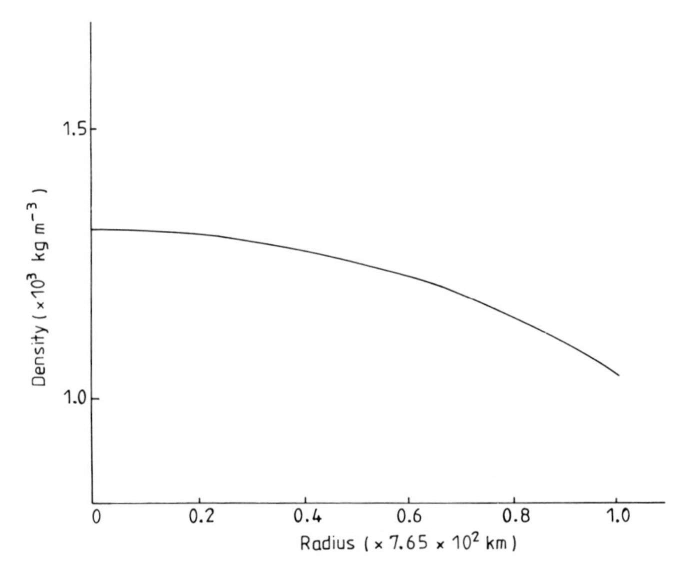

Figure 9.2 *A representative plot of density against depth in a homogeneous ice–silicate satellite with the data adapted to Rhea. Mean density* $= 1.30 \times 10^3$ *kg* m^{-3}*, inertia factor* $= 0.3998$*. (Murnaghan material with* B $= 3$*)*

9.4 THERMAL CONDITIONS

The interiors of the satellites will presumably have been heated during the initial condensation process and any radioactive content will cause heating. The presence of radioactivity can only be surmised at the present time, but it is likely to be there, and surface features show indirect evidence of past heating. It is likely that any radioactive content will be associated with the rock material. For a differentiated body, this means the heat source will be largely in the core and the heat transfer process will be a heat flow through the ice to the surface. Otherwise, for a homogeneous body, the heat sources are spread uniformly throughout the interior. The radioactive content is entirely unknown, but significant conclusions can perhaps be drawn by assuming the concentration to be comparable to that for terrestrial surface rock. In this case, the mean heat production is about 4×10^{-10} W kg^{-1} or some 10^{-2} J kg^{-1} yr^{-1}.

9.4.1 COOLING TIME

According to the theory of the conduction of heat in a solid sphere the time t for initial heat to be lost depends on the radius of the sphere and the thermal diffusivity k of the material according to the expression $t \sim R_0^2/k$. For water ice, $k \sim 10^{-6}$ m^{-2} s^{-1}, while for silicates it is an order of magnitude lower. For an ice sphere of radius 100 km, $t \sim 3 \times 10^8$ years; for a radius 5×10^2 km, $t \sim 7.9 \times 10^9$ years. A sphere of radius 120 km has a cooling time of 4.5×10^9 years, which is the present age of the Solar System. We conclude that the satellites will have retained their initial heat content and that radioactive heating will be adding to the total. Heat is not easily lost by a planetary body, as we saw earlier.

9.4.2 RADIOACTIVE AND RADIATION HEAT BALANCE

To gain some idea of the magnitude of the temperature characteristic of the interior of a planetary body of satellite size, let us consider the equilibrium heat balance that might be approximately established between the possible radioactive decay and radiation from the surface. For terrestrial rock material, the mean heat production is about 4×10^{-10} W kg^{-1} or some 10^{-2} J kg^{-1} yr^{-1}, and we assume this production rate in what follows.

For a differentiated body of the size of Ganymede or Callisto, with a core radius of about 1500 km and mass $M_c = 4 \times 10^{22}$ kg, the heat production H_p will be about 4×10^{20} J yr^{-1} for radioactive heating of terrestrial strength. This quantity of heat would provide the latent heat of fusion (3.4×10^5 J kg^{-1}) for some 10^{15} kg of ice per year at 273 K. If the loss of heat at the surface is neglected, the heat from the core is capable of melting the ice mantle in a year! We have, of course, neglected the loss of heat to the surroundings by radiation through the surface and this is a critical factor in the heat balance. The heat H_r radiated per second from the surface of a body of radius 2×10^6 m (of the order of the size of Ganymede) treated as a black body of unit emissivity is

$$H_r = 4\pi R_0^2 \sigma \langle T \rangle^4$$

where σ ($= 5.67 \times 10^{-8}$ W m^{-2} K^{-4}) is the Stefan radiation constant and $\langle T \rangle$ is the mean surface temperature. The environment has been supposed to be at zero temperature. For a heat balance, this quantity of heat must be the same in magnitude as that radiated, which means

$$1.6\times10^{13}=4\pi R_0^2\sigma\langle T\rangle^4$$

giving $\langle T\rangle=55$ K. This is a low temperature for the surface and would be reduced if the emissivity were lower than unity.

Account must be taken of the absorption of Solar radiation by the surface. At the distance of the Galilean satellites this will be absorbed (for zero surface reflection) at a rate of about 1.2×10^{13} W or about the same as that produced by radioactivity. The surface temperature will now be raised to about 100 K on the sunward side. The difference of temperature between the sunlit and dark hemispheres will be of the order of 50 K. At the distance of Saturn, the solar radiation will have a smaller effect, although the satellite itself will be smaller and the radioactive heat content less.

These estimates are very crude and assume the radioactive content to be that of terrestrial rock, which may not be so. They are, however, interesting as a first estimate of the implications of the models and particularly in indicating the low temperatures to be associated with these icy bodies.

9.4.3 INTERNAL TEMPERATURES

The equilibrium of the planetary body is that for a cold body, and so internal temperatures cannot be deduced from non-thermal observations from outside. The degree of compression is small and can account for only a small temperature rise.

Representative estimates of temperature as a function of the depth, and of the time at a given depth, can be deduced using the general equations of the theory of the conduction of heat in a solid (see chapter 6 and RC6) supplemented by some estimate of the radioactive heat content and distribution in the body. The results suggest a temperature below 300 K even at the centre if the radioactive content is of terrestrial strength, so that the constituent material, if ice, will not truly melt. Solid state convection is a possibility although there is difficulty in constructing a Rayleigh number to explore this possibility. For an initially cold body, calculations show an increase of temperature inside reaching a maximum after some 2×10^9 years after which the body continually cools. Higher temperatures could, of course, be achieved if the internal structure contained non-thermally conducting layers of material or if the surface radiation were reduced due to material peculiarities. This possibility cannot be explored until more is known of the material composition of the bodies.

As a general conclusion for realistic conditions of ice–rock mixtures (though perhaps including as high a proportion as 40% rock) it is found that small bodies (such as Mimas, Enceladus and Hyperion) will remain at very low temperatures throughout their history; intermediate sized bodies (such as

Rhea, Dione, Iapetus and Tethys) show limited melting, although the majority of the volume will remain solid always, but the largest bodies (for instance Titan) can be expected to show considerable melting although the surface will always remain solid. Pluto is classified as a medium sized body in this context, the most likely structure being shown in figure 9.3.

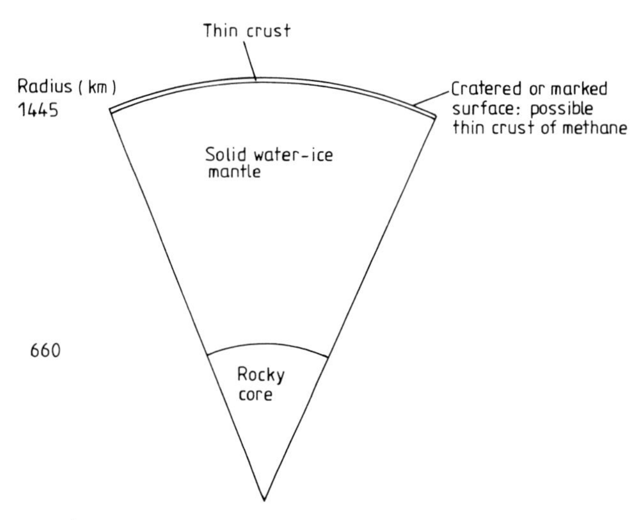

Figure 9.3 *A likely model of the composition of Pluto.*

The same general conclusions will also apply for a homogeneous body where the radioactive heat sources are distributed uniformly throughout the body. The types of ice to be expected in bodies of various sizes is shown in table 9.1.

Table 9.1 *Showing the water-ice forms to be expected in small, medium and large bodies. As examples, Mimas is a small body, Pluto a medium sized one and Callisto a large one.*

Body	Radius ($\times 10^3$ km)	Central pressure ($\times 10^8$ Pa)	Ice forms present
Small	<1	<3	I
Medium	<1.5	<5	I, II
Large	>1.8	>10	I, II, V, VI, VII, VIII

The effects of heating other than by radioactivity may not be negligible for all bodies. For instance, the surface of Enceladus (a small satellite) shows evidence of much evolution and the present contours may not be very old. The only source of heat for such rearrangements can only be tidal in origin, in this case presumably involving Dione.

9.5 COMMENTS ON POSSIBLE DIFFERENTIATION

The third structural possibility mentioned above is for partial differentiation. To assess this possibility it is interesting to gain some idea of the differentiation process under isothermal conditions. The major energy source then will be gravitational.

9.5.1 ENERGY OF DIFFERENTIATION

Differentiation can be supposed to be possible in an initially homogeneous body if the gravitational energy E_d in differentiated form plus the energy dissipated E_v in bringing the differentiation about is greater than the gravitational energy E_u of the homogeneous (undifferentiated) form. To an order of magnitude, for a body of radius R_0, mass M_0 and mean density ρ_0

$$E_u \sim GM_p^2/R_0 = (16\pi^2 G/9)R_0^5\rho_0^2$$

$$E_c \sim (16\pi^2 G/9)R_c^5\rho_0^2$$

$$E_m \sim (16\pi^2 G/9)\rho_m^2(R_0^5 - R_c^5).$$

where ρ_m is the density of the mantle. The subscripts c and m refer to the core and mantle respectively.

The differentiated form is energetically preferred if $E_u > E_d$ which means if $dE = E_u - E_d > 0$. Write $e = E_d/E_u$, so that

$$e = \rho_m^2 + r_1^5(\rho_1^2 - \rho_2^2)$$

where $r_1 = R_c/R_0$, $\rho_1 = \rho_c/\rho_0$ and $\rho_2 = \rho_m/\rho_0$. The differentiation is energetically preferred if $e > 1$ which implies

$$r_1 < [(1-\rho_0^2)/(\rho_1^2 - \rho_2^2)]^{1/5}.$$

This is a weak condition on the size of the core and calculated sizes are smaller than this.

It is convenient to work in terms of the energy associated with each individual atom. The number N_0 of atoms in the body is $N_0 = M_0/Am_p$ where A is the atomic mass number and m_p is the mass of the proton ($=1.67 \times 10^{-27}$ kg). The energy released per atom on differentiation is $de = dE/N_0$.

9.5.2 ENERGY OF BINDING

The atoms in a solid are held together by interatomic forces associated with an energy that can be gauged by the melting temperature T_m. The binding energy

e_b between atoms can be expressed roughly as $e_b = kT_m$ where k ($=1.38 \times 10^{-23}$ J K^{-1}) is the Boltzmann constant. This energy is closely similar to that which is involved with the relative rearrangement of the atoms in space. For differentiation to be possible, we must have $de > e_b$. The mobility of atoms in the satellite can be assessed by forming the ratio $e_f = de/e_b$. The atoms can be regarded as free to differentiate if $e_f > 1$; otherwise they are not.

9.5.3 INTERNAL STRUCTURES

The conditions within the satellites of the major planets are assessed in RC9.4. There it is found that objects of the size of Europa, Ganymede, Callisto, Titan and Triton have ample energy to differentiate from an initially homogeneous form ($e_f > 1$). Objects like Mimas, Enceladus, Tethys, Dione, Iapetus and Nereid do not have sufficient energy ($e_f \ll 1$). This covers the satellites of Jupiter, Saturn and Neptune. The satellites of Uranus are intermediate in that $0.3 < e_f < 0.5$. This could be interpreted as implying no differentiation but, remembering that only mean energies are involved, it could be interpreted as implying that up to 30% of the rock material is differentiated, the remainder being mixed with ice in the now extended mantle. These estimates may well be increased by accounting for other sources of energy; for instance, temperature increases will act towards easier differentiation. Thermal effects could be accounted for, once the radioactive content is known, by using the approach of the thermal pressure (RC8.3) or the Polhäusen approach explained in RC8. The effect of viscosity as a virtual thermostat, explained generally in chapter 5, is still applicable here.

The magnetic conditions are interesting even if the information is scant. There are indications that Titan has an intrinsic magnetic field though the strength is less than 10^{-3} that of the Earth. Io appears to have a field about 10^2 times as large although it is not strictly of interest here because it is not an icy satellite. There is the possibility that Ganymede has a magnetic field of internal origin but the evidence is not conclusive. The existence of such fields must imply a differentiated internal structure and an electrically conducting core, although the implications are far from clear.

The investigation of the ice satellites is still in its infancy but promises interesting developments in the future. Apart from water-ice, interest can be expected to turn to ammonia- and methane-ices and mixtures of these basic types.

9.6 CONCLUSIONS

1 The satellites of Jupiter (with the exception of Io) and of Saturn, as well as the planet Pluto, are known to have water-ice as a major constituent. The same may well be true for the satellites of Uranus and Neptune but little of

a detailed quantitative nature is known of them. The forthcoming Voyager fly-bys will, it is hoped, change this situation.

2 Water-ice is the main rock-forming mineral for these bodies, and methane and ammonia are unlikely to form more than a small impurity, except perhaps in the surface regions, where the temperature is at its lowest.

3 The general properties of ice are not well known although ten polymorphic phases have been recognised and a preliminary phase diagram is available.

4 The internal structures of icy bodies can be approached on the basis of the standard models of a planetary body of given constitution but there is considerable uncertainty in the detailed material composition.

5 The separate cases of full differentiation, partial differentiation and no differentiation at all cannot be distinguished observationally at the present time but each case can be treated analytically.

6 The larger icy bodies (such as Triton or Pluto) are likely to have a differentiated structure of a central rocky core encased in ice, probably largely water-ice. The smaller bodies are likely to be an undifferentiated ice–rock mixture. The bodies of medium size could be undifferentiated or only partially differentiated.

7 The thermal conditions are largely unknown because the radioactive heat content (if any) itself remains unknown. The temperatures of the surface regions are unlikely to be in excess of 10^2 K and are probably less, while the central temperatures could be 3×10^2 K. It is unlikely that the smallest bodies have ever melted internally but the larger ones will have done so below a thin solid crust; some may have liquid mantles now. Convection (perhaps solid state) is likely to have played a significant role in the evolution of the internal structures of the larger bodies.

8 There are indications that Titan has a magnetic field of internal origin (of about 10^{-3} the strength of the Earth's field) and it is possible that Ganymede has such a field as well. If this proves correct the internal structure must be differentiated and presumably there will be an iron-rich core.

9 The study of the icy planetary bodies is still in its infancy but promises exciting discoveries for the future.

REFERENCES AND COMMENTS

RC9.1 PHYSICAL PROPERTIES OF WATER-ICE

The physical properties of the ten polymorphs of ice are not well known. The general features, according to present knowledge, are collected in

Poirier J P 1982 Rheology of ices: a key to the tectonics of the ice moons of Jupiter and Saturn *Nature* **299** 683–8

The effects of adding a sand impurity are described by

Lange M A and Ahrens T J 1981 *EOS* **62** 939
—— 1982 *EOS* **63** 365

A general compendium on ice is

Hogg P V 1975 *Ice Physics* (Oxford: Clarendon)

A very interesting popular description of the peculiar features of water-ice is given by

Asimov I 1976 *The Left Hand of the Electron* (London: Panther) pp 78–104

Ice is not a true solid but is an amorphous material and can be treated as a liquid with very high shear viscosity. It has a higher melting temperature than might be expected on general considerations and this gives a clue to its distinctive spatial molecular features associated with an electric structural dipole. Ammonia (NH_3) also has polar molecules and a melting point of 239.8 K under atmospheric conditions. The boiling point of methane (CH_4) is 111.7 K at atmospheric pressure. The appearance of solid methane and ammonia (ices) within the satellites is therefore restricted very much by internal conditions of temperature. It is very difficult to assess these without a knowledge of the radioactive heat content. The physical properties of ice–methane/ammonia mixtures are not known.

RC9.2 GENERAL CONSIDERATIONS

Photographs of the surfaces of many of the satellites of Jupiter and Saturn are now available and a good collection is in

Briggs G and Taylor F 1982 *The Cambridge Photographic Atlas of the Planets* (Cambridge: Cambridge University Press)

Apart from the excellent photographs themselves there are maps of the surfaces and a very useful commentary on them. Comparable information can ultimately be expected for the satellites of Uranus and Neptune from the Voyager fly-bys over the next decade.

The types of ice to be found at different levels in a satellite will depend primarily on the local pressure. This can be calculated for a body in hydrostatic equilibrium if the distribution of density is known. One way of determining this is to solve equation (3.29) for the mass distribution. Rotation has negligible effect, so we set $\phi_B=0$ and equation (3.29) becomes

$$\frac{\mathrm{d}^2m}{\mathrm{d}r^2}-\frac{2}{r}\frac{\mathrm{d}m}{\mathrm{d}r}+\theta_B r^{(2B-4)}m\left(\frac{\mathrm{d}m}{\mathrm{d}r}\right)^{(2-B)}=0 \qquad \text{(RC9.1)}$$

where

$$\theta_B = (4\pi)^{(B-1)}(G\rho_0^B/K_0)M_0^{(2-B)}R_0^{(3B-4)}.$$

For a homogeneous body with small or moderate compression, the inertia factor α_P is only a few per cent less than 0.4 and m can be expanded as a power series in θ_B, a parameter defined by equation (3.29a), using the boundary conditions (3.30). The detailed analysis is given by

Cole G H A 1971 On Inferring Elastic Properties of the Deep Lunar Interior *Planet. Space Sci.* **19** 929–47

Explicitly we have

$$m(r) = \sum_{j=0}^{\infty} m_j(r)\theta_B^j$$

where

$$m_0(r) = r^3$$

$$m_1(r) = -(3^{(2-B)}/10)(r^5 - r^3)$$

$$m_2(r) = (3^{(3-2B)}/1400)[5(13-5B)r^7 + 42(B-3)r^5 + (61-17B)r^3]$$

and so on. The corresponding expansion for the density is

$$\rho(r)/\rho_P = 1 + \rho_1(r)\theta_B + \rho_2(r)\theta_B^2 + \cdots \qquad \text{(RC9.2)}$$

where

$$\rho_1(r) = (3^{(1-B)}/10)(3-5r^2)$$

$$\rho_2(r) = (3^{2(1-B)}/1400)[3(61-17B) + 210(B-3)r^2 + 35(13-5B)r^4]$$

and so on.

The pressure is similarly expressed as an expansion in the form

$$p(r) = \frac{GM_0\rho_P}{3R_0} \sum_{j=0}^{\infty} p_j(r)\theta_B^j \qquad \text{(RC9.3)}$$

where

$$p_0(r) = \tfrac{3}{2}(1-r^2)$$

$$p_1(r) = (3^{(2-B)}/10)[3(1-r^2) - 2(1-r^4)]$$

$$p_2(r) = (3^{(2-2B)}/4200)[(738-153B)(1-r^2) - (1008-252B)(1-r^4) + (430-125B)(1-r^6)]$$

and so on.

The value of θ_B appropriate to the internal conditions is of the order 10^{-2}, so the magnitude of the pressure is dominated by the leading term. Indeed, for an accuracy of one per cent or less, only the first term of the expansion need be taken into account. Again, the pressure builds up quickly with depth: for $r=\frac{3}{4}$, $p=0.44p_c$ while for $r=\frac{1}{2}$, $p=0.75p_c$ and for $r=\frac{1}{4}$, $p=0.94p_c$. Good preliminary estimates of the pressure distribution within the body can therefore be built up using very elementary formulae.

The central pressures derived from these formulae for the satellite bodies are set out in table RC9.1 for Jupiter and Saturn and in table RC9.2 for Uranus and Neptune.

Table RC9.1 *The values of the central pressures for the satellites of Jupiter and Saturn assuming them to be of homogeneous composition and using the formula (RC9.3) for the pressure.*

Body	Central pressure ($\times 10^6$ N m^{-2})	Ice forms	Body	Central pressure ($\times 10^6$ N m^{-2})	Ice forms
Europa	3161	I, II, VI, VII	Mimas	7.79	I
Ganymede	3635	I, II, VI, VII	Enceladus	11.1	I
Callisto	2665	I, II, VI, VII	Tethys	41.1	I
Titan	33638	I, II, VI, VII	Dione	89.2	I
Rhea	120.9	I	Iapetus	110.3	I

Table RC9.2 *The values of the central pressures for the satellites of Uranus and Neptune together with the possible associated ice forms. Use is made of equation (RC9.2)*

Body	Central pressure ($\times 10^6$ N m^{-2})	Ice forms
Miranda	15.75	I
Ariel	105.0	I
Umbriel	112.0	I
Titania	156.8	I
Oberon	119.5	I
Triton	1384.0	I, II, V, VI, VII
Nereid	13.2	I

It is seen that a variety of ice forms can be expected in the central regions of the larger satellites but only ice I is of significance for the smaller bodies. The same conclusions apply for a differentiated body. The pressure is reduced by no more than 25% even midway to the centre, so a core structure will have little practical effect on the ice forms present. The central pressure will, of course, be increased by the differentiation of material but again by only a few per cent for bodies of satellite size.

RC9.3 MODEL SATELLITES

For a body with prescribed mean density ρ_P and surface density $\rho(1)$, equation (RC9.2) allows the corresponding value of θ_B to be calculated. Because R_0, M_0 and ρ_0 are known, equation (3.29a) then gives a value for the zero pressure bulk modulus K_0.

For pure water-ice we take $B=3$ and $\rho(1)=920$ kg m^{-3}. The value of K_0 is known then to be about 1×10^{10} N m^{-2}. The addition of an impurity of rock material (with density at zero pressure about 3500 kg m^{-3}, although the precise value is open to some uncertainty) will increase the values of the density and bulk modulus. For a proportion x by mass of rock the density is

$$\rho(\text{rock/ice})=x\rho(\text{rock})+(1-x)\rho(\text{ice})$$

and the corresponding K_0 is obtained from equation (3.29a). The dependence of K_0 on the proportion x of rock for a body of given mass and radius is obtained by repeating this calculation for different values of x. The results for the satellites of the major planets are contained in table RC9.3. Nereid can be disregarded from this point onwards as being too small to become internally differentiated.

Table RC9.3 *Giving the calculated proportions and masses of rock in the ice satellites of the major planets and for Pluto using equation (RC9.3) for the density. The density of the rock at zero pressure is assumed to be 3500 kg m^{-3} and for water-ice the density is 920 kg m^{-3}.*

Body	Proportion of rock, x (%)	Mass of rock ($\times10^{19}$ kg)	Body	Proportion of rock, x (%)	Mass of rock ($\times10^{19}$ kg)
Europa	80.6	3928	Rhea	10.7	24.4
Ganymede	38.0	5662	Iapetus	11.5	22.2
Callisto	33.0	3511	Miranda	45.7	1.64
Titan	37.5	5095	Ariel	42.3	28.3
Pluto	13.5	210.6	Umbriel	41.5	31.5
Mimas	10.5	0.39	Titania	43.3	52.0
Enceladus	7.4	0.55	Oberon	42.3	34.7
Tethys	3.6	2.25	Triton	41.0	2337.0
Dione	19.0	19.95			

If this rock were concentrated at the centre to form a core it would have the radii listed in table RC9.4. Whether such a differentiated form is to be expected in practice requires a separate discussion.

Table RC9.4 *Calculated radii for the rock cores of the principal satellites of the major planets assuming the material described in table RC9.3 is collected at the centre.*

Body	Core radius R_c ($\times 10^2$ km)	R_c/R_0	Body	Core radius R_c ($\times 10^2$ km)	R_c/R_0
Europa	14.0	0.88	Rhea	2.55	0.33
Ganymede	15.0	0.59	Iapetus	2.47	0.34
Callisto	13.0	0.55	Miranda	1.04	0.64
Titan	15.0	0.59	Ariel	2.68	0.62
Pluto	6.5	0.44	Umbriel	2.78	0.62
Mimas	0.65	0.33	Titania	3.29	0.63
Enceladus	0.72	0.29	Oberon	2.87	0.62
Tethys	1.15	0.22	Triton	11.7	0.61
Dione	2.39	0.43			

The indications from these calculations are that the internal structures of the satellites of Uranus are somewhat different from those of the other major planets. It may be that the observational data of these objects are not of sufficient accuracy at the present time, although the data for the two satellites of Neptune lead to comparable structures to those of Jupiter and Saturn.

The non-silicate material has been supposed to be water-ice and no account has been taken of methane or ammonia impurities. The effects of these on the physical properties of ice are not known.

RC9.4 COMMENTS ON POSSIBLE DIFFERENTIATION

We consider first the differentiation of the rock and ice, and this is followed by the more hypothetical question of the differentiation of any core that might be formed.

RC9.4.1 Differentiation into mantle and core. The formulae of the main text are applied to the study of the satellites of the major planets by using the physical data of tables 1.3 and 1.4 together with the zero pressure density for rock as 3500 kg m^{-3} and of ice I as 920 kg m^{-3}. Higher density polymorphs of ice can be neglected to the present accuracy because they will appear only deep in the interior and will not affect the mean density of the ice critically. The ratios $e_f = de/e_b$ are given in table RC9.5.

The binding energy will be that appropriate to the major component. For a substantial rock component, we take the energy appropriate to iron as representative and $e_b = 3 \times 10^{-20}$ J per atom; for ice $e_b = 5 \times 10^{-21}$ J per atom.

Table RC9.5 *The ratios of the gravitational energy per atom in the satellite to an average atomic binding energy per atom. Differentiation can be expected for* $e_f > 1$.

Body	e_f	Body	e_f	Body	e_f
Europa	1.5	Dione	0.33	Mimas	0.02
Ganymede	11.9	Rhea	0.34	Enceladus	0.03
Callisto	9.1	Iapetus	0.33	Tethys	0.05
Titan	11.7	Ariel	0.43	Miranda	0.06
Triton	6.8	Umbriel	0.46		
Pluto	1.15	Oberon	0.49		

On the basis of table RC9.5 we can conclude that the larger bodies listed in the first column can all be expected to be differentiated. The bodies listed in the fifth column are unlikely to be differentiated. Those listed in the third column present a separate problem. They fall short of the energy needed by an amount that could well be within the errors of our calculations. Alternatively, they might represent partial differentiation in which a core is formed while some material remains suspended in ice. If this were so, nearly a half of the rock component would be separated out, the remainder being in the mantle. The interpretation involving partial differentiation would be within the spirit of mean energies used in the calculations.

It is necessary to account for energies of dissipation associated with the differentiation process, and here the viscous energy of motion is the essential one. We will assume that the material can be regarded as a Newtonian liquid of very high shear viscosity, with coefficient μ, as far as the discussion of instabilities and the beginning of motion are concerned. Then the viscous energy E_v for motion with characteristic speed U in a region of characteristic size L is, to an order of magnitude

$$E_v \sim \mu(U/L)^2.$$

For differentiation to occur, $E_u - E_d = dE \gg E_v$. The speed U can be estimated by realising that the maximum motion occurs when a steady state of balance is reached between the gravitational and frictional forces. For a spherical body of mass m and radius a the gravitational force is mg, while the viscous force is described by Stokes's law; this means

$$mg \sim 6\pi a \mu U$$

from which we see that

$$U \sim mg/6\pi a \mu.$$

The large shear viscosity ensures that the speeds will always be small except for the very smallest particles. The value of the shear viscosity coefficient for ice is of the order 10^{13} N m^{-2}, from which it follows that a ferrous sphere of radius 1.4×10^{-1} m would require 4.5×10^9 years to fall from the surface to the centre of a body of radius 10^6 m. Smaller spheres would take longer, and larger spheres would fall more quickly. The differentiation process for an initially homogeneous body could therefore proceed for the larger spheres but the smaller members of a distribution of sizes would not yet have had time to become differentiated. A range of initial sizes of particle would, on this basis, show a partial differentiation, and such a possibility was deduced earlier on different arguments. We are here becoming involved with conditions in the very early Solar System.

RC9.4.2 Differentiation of the core itself. If it is supposed that the rock material contains a silicate component and, separately, an iron component as a free metal, the arguments we have just used can be applied again to consider the possibility of the core separating into a silicate outer core and a ferrous inner core. We consider only the bodies listed in the first column of table RC9.5 because it is not clear that the others will be differentiated at all.

Table RC9.6 *The ratio of the energies of gravitational separation and binding (per atom) for the central rocky cores of the six major ice bodies of the outer Solar System.* $\langle\langle T\rangle\rangle$ *are the corresponding associated mean temperatures.*

Body	Iron core radius ($\times 10^5$ km)	e_f	$\langle\langle T\rangle\rangle$ ($\times 10^3$ K)
Europa	4.85	0.99	2.1
Ganymede	7.72	2.60	5.7
Callisto	6.19	1.64	3.6
Titan	6.23	1.58	3.4
Triton	4.50	0.79	1.7
Pluto	2.07	0.17	0.37

The ratio e_f = (energy of differentiation)/(energy of binding), used before, has the values listed in table RC9.6. As before, differentiation can be expected for $e_f > 1$. It is seen from the table that this condition applies for Europa, Ganymede, Callisto and Titan, but not for Pluto or Triton. The gravitational energy is associated with a mean temperature $\langle\langle T\rangle\rangle = e_0/k$ (k is the Boltzmann constant). This can be interpreted as indicative of the mean temperature of the region, but is not to be accepted as an actual temperature. Again, more precise binding energies would be needed if the calculations were to be made more accurate. It is seen from the table that, with the exception of Pluto, the calculated temperatures are of the order of 10^3 K. The time taken for differentiation to occur in the rocky core will be larger than for ice because the viscosity is also larger—probably by a factor of about 10^5 if temperature is also taken into account. There could well be problems of time scale here, the differentiation process not yet having had time to become complete. Information on the degree of differentiation (if any) of these bodies could well have a bearing on the conditions of formation of the Solar System. For comments on these problems, see

Cole G H A 1983 Interiors of the Icy Satellites of Saturn *The Observatory* **103** 293–6
—— 1984 Interior Structure of the Icy Satellites and of Pluto *Q. J. R. Astron. Soc.* **25** 19–27

The presence of an iron core would have more immediate implications. According to the arguments of chapter 6, the physical situations in the cores of Europa, Ganymede, Callisto and Titan could be suitable for the production of magnetic fields in these bodies. Evidence for such a field has, in fact, been found for Titan and it is likely that Ganymede has a field as well. Further details can be found in

Stevenson D J 1983 Planetary Magnetic Fields *Rep. Prog. Phys.* **46** 555–620

The actual production of the magnetic field could require finer constraints on the system and need not be found to apply for all the satellite bodies. The presence of magnetic fields with sources within these planetary bodies would have important theoretical consequences.

Index